PLANT SCIENCE RESEARCH AND PRACTICES

OCIMUM BASILICUM

TAXONOMY, CULTIVATION AND USE

PLANT SCIENCE RESEARCH AND PRACTICES

Additional books and e-books in this series can be found on Nova's website under the Series tab.

PLANT SCIENCE RESEARCH AND PRACTICES

OCIMUM BASILICUM

TAXONOMY, CULTIVATION AND USE

ANDRES A. WALTON
EDITOR

Copyright © 2021 by Nova Science Publishers, Inc.

All rights reserved. No part of this book may be reproduced, stored in a retrieval system or transmitted in any form or by any means: electronic, electrostatic, magnetic, tape, mechanical photocopying, recording or otherwise without the written permission of the Publisher.

We have partnered with Copyright Clearance Center to make it easy for you to obtain permissions to reuse content from this publication. Simply navigate to this publication's page on Nova's website and locate the "Get Permission" button below the title description. This button is linked directly to the title's permission page on copyright.com. Alternatively, you can visit copyright.com and search by title, ISBN, or ISSN.

For further questions about using the service on copyright.com, please contact:
Copyright Clearance Center
Phone: +1-(978) 750-8400   Fax: +1-(978) 750-4470   E-mail: info@copyright.com.

NOTICE TO THE READER

The Publisher has taken reasonable care in the preparation of this book, but makes no expressed or implied warranty of any kind and assumes no responsibility for any errors or omissions. No liability is assumed for incidental or consequential damages in connection with or arising out of information contained in this book. The Publisher shall not be liable for any special, consequential, or exemplary damages resulting, in whole or in part, from the readers' use of, or reliance upon, this material. Any parts of this book based on government reports are so indicated and copyright is claimed for those parts to the extent applicable to compilations of such works.

Independent verification should be sought for any data, advice or recommendations contained in this book. In addition, no responsibility is assumed by the Publisher for any injury and/or damage to persons or property arising from any methods, products, instructions, ideas or otherwise contained in this publication.

This publication is designed to provide accurate and authoritative information with regard to the subject matter covered herein. It is sold with the clear understanding that the Publisher is not engaged in rendering legal or any other professional services. If legal or any other expert assistance is required, the services of a competent person should be sought. FROM A DECLARATION OF PARTICIPANTS JOINTLY ADOPTED BY A COMMITTEE OF THE AMERICAN BAR ASSOCIATION AND A COMMITTEE OF PUBLISHERS.

Additional color graphics may be available in the e-book version of this book.

Library of Congress Cataloging-in-Publication Data

ISBN: 978-1-53619-265-0

Published by Nova Science Publishers, Inc. † New York

CONTENTS

Preface vii

Chapter 1 Secondary Metabolites of *Ocimum bacilicum* L. 1
Sıbel Day

Chapter 2 Sustainable Approaches in
Ocimum basilicum Cultivation 29
Baraa Almansour

Chapter 3 An Overview of *O. Basilicum* (L.) in Turkey 51
Muhammad Azhar Nadeem, Yeter Çilesiz,
Ecenur Korkmaz, Zemran Mustafa,
Faheem Shehzad Baloch, Tolga Karaköy and
Muhammad Aasim

Chapter 4	*Ocimum basilicum* as a Potential Anti-COVID-19 Plant: Review on the Antiviral Activity and Molecular Docking of Some of Its Molecules with the SARS-Cov-2 Main Protease (MPRO) *Pius T. Mpiana, Etienne M. Ngoy, Jason T. Kilembe, Carlos N. Kabengele, Aristote Matondo, Clement L. Inkoto, Emmanuel M. Lengbiye, Domaine T. Mwanangombo, Damien S. T. Tshibangu, Koto-te-Nyiwa Ngbolua and Dorothée D. Tshilanda*	73
Chapter 5	Antisickling Activity of *Ocimum basilicum* and Some of Its Compounds *Dorothée D. Tshilanda, Carlos N. Kabengele, Etienne M. Ngoyi, Aristote Matondo, Jason T. Kilembe, Giresse N. Kasiama, Clement L. Inkoto, Emmanuel M. Legbiye, Benjamin Z. Gbolo, Gédéon N. Bongo, Damien S. T. Tshibangu, Koto-te-Nyiwa Ngbolua and Pius T. Mpiana*	113
Index		135

PREFACE

Ocimum bacilicum L. is an aromatic herb commonly known as sweet basil or sweet tulsi. It is rich in secondary metabolites like phenols, alkaloids, terpenoids, aldehydes, flavonoids, steroids, glycosides, essential oils, saponins, and tannins. The presence of these compounds makes sweet basil one of the most commonly used plant in aromatherapy, perfume, cosmetics, and in foods. The utilization potential of the sweet basil in different industrial section increases its importance. The first chapter underlines secondary metabolites of sweet basil and their importance in different aspects. The second chapter considers the recent concepts of application organic manures in integration with inorganic fertilizers in different reviews and research studies that fulfills the nutritional needs in sweet basil and gives the best quality of it. The third chapter summarizes the potential uses, cultivation, and available germplasm of *O. basilicum* in Turkey. The fourth chapter reviews literature on antiviral activity of *O. basilicum* to find molecules capable of inhibiting the SARS-CoV-2 main protease. This could permit the use of this plant in the fight against COVID-19 and associated diseases. The last chapter is an examination of antisickling activity of *Ocimum Basilicum* and some of its compounds.

Chapter 1 - *Ocimum bacilicum* L. is an aromatic herb commonly known as sweet basil or sweet tulsi. It is rich in secondary metabolites like phenols, alkaloids, terpenoids, aldehydes, flavonoids, steroids, glycosides, essential oils, saponins, and tannins. The presence of these compounds makes sweet basil one of the most commonly used plant in aromatherapy, perfume, cosmetics, and in foods. The utilization potential of the sweet basil in different industrial section increases its importance. This paper will underline secondary metabolites of sweet basil and their importance in different aspects.

Chapter 2 - The plant kingdom a vast storehouse for several species is a source for chemical molecules waiting to be identified, isolated, manipulated and used. Medicinal and aromatic plants constitute a major segment of the flora reputed to possess distinct alkaloids and fragrances that are considered as a source for therapeutic agents in health care. The past three decades have obviously witnessed a tremendous surge in acceptance and public interest in natural therapies both in developing and developed countries. It is estimated that, up to four billion people (representing 85% of the world's population) living in the developing world rely on herbal medicinal products as a primary source of traditional medical practice. Moreover, 120 drugs in western medicine are obtained from plants, while many other drugs are obtained either by semi synthesis from plant products, or synthesis based on plant molecules. Among of these species of medicinal crops is *Ocimum basilicum* L. plant which has the most economic importance and is well known as sweet basil/French basil. It has diversified use in perfumery, pharmaceutical industries, confectionery as well as in food, flavor, dental, oral products, and traditional medicines, so the demand for it is growing in recent years. As the knowledge and awareness about global environmental issue, integrated supply of nutrients to plants through planned combinations of organic and inorganic sources is becoming an increasingly important aspect of environmentally sound agriculture. Taking in considering that the real value is given to the quality in cultivation of Ocimum, while yield quantity comes in the second step of importance, this chapter will consider the recent concepts of application organic manures in integration with inorganic fertilizers in different reviews and research studies that fulfills the nutritional needs in sweet basil and gives the best quality of it.

Chapter 3 - *Ocimum basillicum L. sweet basil is an annual herbaceous plant belonging to Ocimum genus of the Lamiaceae family.* The genus *Ocimum* contains 50 to 150 herb and shrub species and shows the natural distribution in Asia, Africa, and Central America. Sweet basil has a good concentration of vitamins and minerals and potentially used as a medicinal plant for the treatment of various diseases. Sweet basil reflects great variations in terms of its morphology and chemical contents and is mainly grown as a spice and ornamental plant. The essential oil contents of sweet basil vary 0.5% to 1.5% depending on the climatic conditions. Additionally, sweet basil is considered a good source of various phenolic compounds that have beneficial effects on human health. Sweet basil is also used as spice, medicine, food, and perfumery industries. The essential oil of this plant has increasing importance due to various biological effects such as antifungal, insecticide, and antioxidant. In addition, the purple-colored varieties of sweet basil are an important anthocyanin source for the food industry. In Turkey, sweet basil is widely used for culinary and medicinal purposes and its leaves are used as salad. There is a number of registered cultivars of sweet basil in Turkey that are used for different purposes ranging from food to medicine and ornamental plant. This study summarized the potential uses, cultivation, and available germplasm of *O. basilicum* in Turkey.

Chapter 4 - Coronavirus disease, a pandemic that has already caused more than one million deaths worldwide, has not yet found an effective and safe treatment; hence the need to resort to medicinal plants. *Ocimum basilicum* is an edible plant that has also shown several biological properties including antiviral, antisickling, antioxidant, anti-inflammatory, etc. Data collected in the literature have shown that the molecules contained in *O. basilicum* possess antiviral properties against several viruses (Herpes Simplex virus-1 and 2, Human Immunodefiency Virus-1, Adeno virus, Hepatitis B Virus, Enterovirus 71, Coxsackie virus B1, etc.). The molecular docking carried out between some molecules of this plant with the main protease (Mpro) of SARS-CoV-2 involved in the replication of the virus responsible for COVID-19 shows interesting interactions and stables complexes. These molecules could act either alone or in synergy to inhibit viral replication. In addition, aromatic plants including *O. basilicum* are used in Congolese traditional medicine for the treatment of respiratory and inflammatory diseases. The objective of this work is therefore to review the literature on the antiviral activity of *O. basilicum* and to find molecules capable of inhibiting the SARS-CoV-2 main protease. This could permit the use of this plant in the fight against COVID-19 and associated diseases.

Chapter 5 - Sickle cell disease is an inherited disorder characterized by a structural abnormality of hemoglobin S. According to the WHO, several thousand people worldwide suffer from this disease, with the majority of cases recorded in Africa. The ineffectiveness and inaccessibility of the treatments presented, force populations to resort to medicinal plants. Thus, in addition to its culinary use as a spice, *Ocimum basilicum* is also cited in the treatment of various diseases such as cancer, diarrhea, hemorrhoids, rhinitis including sickle cell disease. The antisickling activity of the acidified methanol extract, anthocyanin fraction and essential oil as well as that of butyl stearate and rosmarinic acid two pure compounds isolated from Ocimum basilicum L was evaluated using the Emmel test. The results showed that all these extracts and pure compound exhibited an interesting antisickling activity. These activities are dose-dependent with normalization rate higher than 80%. Rosmarinic acid and methanolic extract showed a minimum normalization concentrations of 0.18 ± 0.03 mg/mL and 0.23 ± 0.04 mg/mL respectively. Essential oil showed good antioxidant activity with $IC_{50} = 1.180 \pm 0.015$ using the DPPH test.

In: *Ocimum basilicum*
Editor: Andres A. Walton

ISBN: 978-1-53619-265-0
© 2021 Nova Science Publishers, Inc.

Chapter 1

SECONDARY METABOLITES OF *OCIMUM BACILICUM* L.

Sibel Day[*]
Department of Field Crops, Faculty of Agriculture,
Ankara University, Ankara, Turkey

ABSTRACT

Ocimum bacilicum L. is an aromatic herb commonly known as sweet basil or sweet tulsi. It is rich in secondary metabolites like phenols, alkaloids, terpenoids, aldehydes, flavonoids, steroids, glycosides, essential oils, saponins, and tannins. The presence of these compounds makes sweet basil one of the most commonly used plant in aromatherapy, perfume, cosmetics, and in foods. The utilization potential of the sweet basil in different industrial section increases its importance. This paper will underline secondary metabolites of sweet basil and their importance in different aspects.

---

[*] Corresponding Author's E-mail: day@ankara.edu.tr.

INTRODUCTION

Ocimum basilicum L. belonging to family of Lamiaceae is an annual or short-lived perennial herb commonly cultivated around the world (Harnafi et al. 2013). Sweet basil is mostly used as ornamental plant and the aerial part of the plant is used in food as a flavoring agent, medicine and cosmetology (Burdina and Priss 2016). Its leaves and flowers are known as carminative, galactogogue, stomachic and antispasmotic in folk medicine (Sajjadi, 2006).

The quantity and quality of secondary metabolites in sweet basil can vary within wide ranges and depend on cultivation methods, climate conditions and basil cultivar. Sweet basil is cultivated in many countries, especially in Mediterranean region. Depending on cultivating in different countries and regions it has wide range in secondary metabolites compounds. The aromatic variety it has mainly depends on the major chemical compounds of volatile oil. The major component in volatile oil determines the chemotype of the sweet basil. Compounds like linalool, eugenol, methylchavicol, methylcinnamate and citral are found in chemotypes of basil.

Volatile oil of sweet basil and its component have the potential of insecticidal effects and disrupt insect growth at several life stages. The plant also have carminative, stimulant, diaphoretic, diuretic, dyspepsia, antiseptic, anesthetic, flatulence, gastritis, anti-spasmodic, anthelmintic, anti-diarrheal, analgesic and anti-tussive characteracteristics. Considering its characteristics mentioned before, sweet basil has an important role not only in traditional medicine but also in modern pharmacology and clinical investigations (Shirazi 2014).

Sweet basil volatile oil is rich in chemical components and usage of volatile oil in different industrial and environmental products is increasing. Its antimicrobial, insecticidal effects could be used for saving foods during storage and also could be effective in greenhouse and field pests. Farmers are in trouble when their crops in the field or greenhouse under the attack of diseases or pests during vegetation and after harvesting. Crop production and yield most of the time depends on synthetic chemicals due to their protection effects against disease and pests. However synthetic chemicals utilized in agricultural are risk factors in many ways such as diseases in human being because of residues. They could also lead to the death of birds and beneficial insects playing a role as pollinators, natural enemies, weed killers, soil builders, scavengers (feed on the dead and decaying matter of plants and animals) by overusing. As a result of pesticide risks, there is an increasing demand for environmental saving products in the agricultural production cycle. This tremendous trend in using organic chemicals instead of synthetic ones leading to standardized product commercialization of carrying minimum hazardous effect and searching for secondary metabolites of plants like sweet basil.

In this review some of the important impact of *O. basilicum* secondary metabolites are discussed to shed light on their usage.

CHEMICAL COMPOSITION OF *OCIMUM BASILICUM* VOLATILE OIL

Several studies has been conducted in various part of the world about chemical composition of basil volatile oil. Diversity between cultivars about chemical components of the volatile oil has also been reported. Variety in volatile oil of *Ocimum basilicum* depends on leaf, flower colors aroma and plant origin (Da silva et al. 2003, Sajjadi 2006). Depending on variety in volatile oil components four major chemotypes are mainly known and these are methyl chavicol (estragole) rich, linalool rich, methyl eugenol rich, methyl cinnamate rich (Lawrence et al. 1988). There has been also more than 200 chemical component determined in volatile oil of basil cultivated around the world. Some of the chemical components of basil volatile oil reported by researchers are presented in Table 1.

Table 1. Chemical composition of sweet basil volatile oil

Compounds	Molecular Formula	Reference
Terpenes		
Monoterpene hydrocarbons		
α-Pinene	$C_{10}H_{16}$ *	Baldim et al. 2018; Beatovic et al. 2015; Hadush et al. 2015; Ismail M 2008
α-Phellandrene	$C_{10}H_{16}$	Hassanpouraghdam et al. 2010
α-Terpinolene	$C_{10}H_{16}$	Chang et al. 2009; Liber et al. 2011
α- Terpinene	$C_{10}H_{16}$	Baldim et al. 2018; Beatovic et al. 2015
α-Thujene	$C_{10}H_{16}$	Beatovic et al. 2015; Padalia et al. 2013; Özcan and Chalchat
β-Pinene	$C_{10}H_{16}$	Baldim et al. 2018; Beatovic et al. 2015; Hassanpouraghdam et al. 2010
β-Myrcene	$C_{10}H_{16}$	Chang et al. 2009; Ismail M 2008; Hassanpouraghdam et al. 2010
β- Phellandrene	$C_{10}H_{16}$	Abdallah et al. 2017; Özcan and Chalchat 2002
β-Ocimene	$C_{10}H_{16}$	Chang et al. 2009
δ-3-Carene	$C_{10}H_{16}$	Chang et al. 2009
γ-Terpinene	$C_{10}H_{16}$	Baldim et al. 2018; Beatovic et al. 2015; Hassanpouraghdam et al. 2010
cis-β-Ocimene	$C_{10}H_{16}$	Hussain et al. 2008; Hassanpouraghdam et al. 2010
trans-β-Ocimene ((E)-β-Ocimene; β-Ocimene)	$C_{10}H_{16}$	De Martino et al. 2009; Liber et al. 2011
cis-Sabinene hydrate	$C_{10}H_{18}$	Baldim et al. 2018; Beatovic et al. 2015
Azulene	$C_{10}H_{18}$	Ismail M 2006
Camphene	$C_{10}H_{16}$	Baldim et al. 2018; Beatovic et al. 2015; Ismail M 2006
Limonene	$C_{10}H_{16}$	Baldim et al. 2018; Beatovic et al. 2015; Toncer et al. 2017; Özcan and Chalchat 2002

Compounds	Molecular Formula	Reference
Myrcene	$C_{10}H_{16}$	Baldim et al. 2018; Beatovic et al. 2015; Liber et al. 2011
Ocimene (Trans-β-Ocimene; €-β-Ocimene; β-Ocimene	$C_{10}H_{16}$	Beatovic et al. 2015; Ismail M 2006
P-mentha-1(7),8 diene	$C_{10}H_{16}$	Hadush et al. 2015
Sabinene	$C_{10}H_{16}$	Baldim et al. 2018; Beatovic et al. 2015; Padalia et al. 2013
Terpinolene	$C_{10}H_{16}$	Baldim et al. 2018; Toncer et al. 2017
Oxygenated Monoterpenes		
α-Fenchyl acetate	$C_{10}H_{16}O_2$	Abdallah et al. 2017
α-Terpineol	$C_{10}H_{18}O$	Baldim et al. 2018; Beatovic et al. 2015; Padalia et al. 2013
δ-terpineol	$C_{10}H_{18}O$	Beatovic et al. 2015
p-Cymene	$C_{10}H_{16}O$	Beatovic et al. 2015; Hassanpouraghdam et al. 2010; Özcan and Chalchat 2002
cis-Ocimene	$C_{10}H_{14}O$	Baldim et al. 2018
cis-Linalool oxide	$C_{10}H_{18}O_2$	Beatovic et al. 2015; Liber et al. 2011; Özcan and Chalchat 2002
cis-Rose oxide	$C_{10}H_{20}O$	Hassanpouraghdam et al. 2010
Exo-2-hydroxycineole-acetate	$C_{12}H_{20}O_3$	Hassanpouraghdam et al. 2010
trans-Linalool oxide	$C_{10}H_{18}O_2$	Liber et al. 2011; Özcan and Chalchat 2002
trans-Myroxide	$C_{10}H_{16}O$	Beatovic et al. 2015; Özcan and Chalchat 2002
1,8-Cineole	$C_{10}H_{18}O$	Beatovic et al. 2015; Sajjadi 2006; Ismail M 2006
Borneol	$C_{10}H_{18}O$	Beatovic et al. 2015; Toncer et al. 2017; Padalia et al. 2013
Bornyl acetate	$C_{12}H_{20}O_2$	Baldim et al. 2018; Beatovic et al. 2015; Liber et al. 2011
Camphor	$C_{10}H_{16}O$	Baldim et al. 2018; Beatovic et al. 2015; Ismail M 2006
Carvacrol	$C_{10}H_{14}O$	Toncer et al. 2017; Hassanpouraghdam et al. 2010; Özcan and Chalchat 2002
Carvone	$C_{10}H_{14}O$	Maggio et al. 2016
Citronellal	$C_{10}H_{18}O$	Padalia et al. 2016; Rodrigez-González et al. 2019

Table 1. (Continued)

Compounds	Molecular Formula	Reference
Citronellol	$C_{10}H_{20}O$	Padalia et al. 2013
Endo-Fenchol	$C_{10}H_{18}O$	Liber et al. 2011
Estragol	$C_{10}H_{12}O$	Baldim et al. 2018
Eugenol	$C_{10}H_{12}O_2$	Baldim et al. 2018; Ismail M 2006; Toncer et al. 2017
Fenchol	$C_{10}H_{18}O$	Baldim et al. 2018; Liber et al. 2011
Fenchone	$C_{10}H_{16}O$	Baldim et al. 2018; Beatovic et al. 2015; Hadush et al. 2015
Fenchyl acetate	$C_{12}H_{20}O_2$	Baldim et al. 2018; Beatovic et al. 2015; Liber et al. 2011
Geranial	$C_{10}H_{16}O$	Beatovic et al. 2015; Sajjadi 2006; Padalia et al. 2013; Özcan and Chalchat 2002
Geraniol	$C_{10}H_{18}O$	Maggio et al. 2016; Padalia et al. 2013; Özcan and Chalchat 2002
Geranyl acetate	$C_{12}H_{20}O_2$	Padalia et al. 2013; Liber et al. 2011; Özcan and Chalchat 2002
Iso-neomenthol	$C_{10}H_{20}O$	Hassanpouraghdam et al. 2010
Isopulegol acetate	$C_{12}H_{20}O_2$	Hassanpouraghdam et al. 2010
Linalool	$C_{10}H_{18}O$	Baldim et al. 2018; Beatovic et al. 2015; Ismail M 2006
Linalool oxide	$C_{10}H_{18}O_2$	Baldim et al. 2018
Linalyl acetate	$C_{12}H_{20}O_2$	Hussain et al. 2008; Dris et al. 2017
Menthol	$C_{10}H_{20}O$	Hassanpouraghdam et al. 2010
Menthone	$C_{10}H_{18}O$	Hassanpouraghdam et al. 2010
Methyl chavicol (Estragole)	$C_{10}H_{12}O$	Beatovic et al. 2015; Hadush et al. 2015; Sajjadi 2006; Maggio et al. 2016; Liber et al. 2011
Myrtenol	$C_{10}H_{16}O$	Ismail M 2006; Padalia et al. 2013
Neral	$C_{10}H_{14}O$	Beatovic et al. 2015; Sajjadi 2006; Padalia et al. 2013
Nerol	$C_{10}H_{18}O$	Beatovic et al. 2015; Padalia et al. 2013

Compounds	Molecular Formula	Reference
Neryl acetate	$C_{12}H_{20}O_2$	Padalia et al. 2013; Liber et al. 2011
Octan-3-yl acetate	$C_{10}H_{20}O_2$	Özcan and Chalchat 2002
Octanol acetate	$C_{10}H_{20}O_2$	Beatovic et al. 2015
Pinocarvone	$C_{10}H_{14}O$	Özcan and Chalchat 2002
Piperitone	$C_{10}H_{16}O$	Hassanpouraghdam et al. 2010
Pulegone	$C_{10}H_{16}O$	Hassanpouraghdam et al. 2010
Terpinen-4-ol	$C_{10}H_{18}O$	Baldim et al. 2018; Beatovic et al. 2015; Sajjadi 2006
Thymol	$C_{10}H_{16}O$	Maggio et al. 2016; Liber et al. 2011
4-acetyl-1-methylcyclohexene	$C_9H_{14}O$	Beatovic et al. 2015
Sesquiterpene hydrocarbons		
α-Acoradiene	$C_{15}H_{24}$	Özcan and Chalchat 2002
α-Amorphene	$C_{15}H_{24}$	Hassanpouraghdam et al. 2010; Liber et al. 2011
α-Bergamotene	$C_{15}H_{24}$	Hussain et al. 2008; Liber et al. 2011
α-trans-Bergamotene	$C_{15}H_{24}$	Beatovic et al. 2015; Padalia et al. 2013
α-Bisabolene	$C_{15}H_{24}$	Liber et al. 2011
α-Bulnesene (δ-guaiene)	$C_{15}H_{24}$	Baldim et al. 2018; Beatovic et al. 2015; Padalia et al. 2013
α-Cadinene	$C_{15}H_{24}$	Baldim et al. 2018; Padalia et al. 2013
α-Copaene	$C_{15}H_{24}$	Baldim et al. 2018; Hassanpouraghdam et al. 2010
α-Cubebene	$C_{15}H_{24}$	Hadush et al. 2015; Ismail M 2006; Hassanpouraghdam et al. 2010
α-Farnesene	$C_{15}H_{24}$	Liber et al. 2011
α-Guaiene	$C_{15}H_{24}$	Baldim et al. 2018; Beatovic et al. 2015; Hassanpouraghdam et al. 2010
α-Gurjunene	$C_{15}H_{24}$	Toncer et al. 2017; Padalia et al. 2013
α-Muurolene	$C_{15}H_{24}$	Padalia et al. 2013
α-Selinene	$C_{15}H_{24}$	Baldim et al. 2018; Padalia et al. 2013
β-Acoradiene	$C_{15}H_{24}$	Özcan and Chalchat 2002

Table 1. (Continued)

Compounds	Molecular Formula	Reference
β-Bisabolene	$C_{15}H_{24}$	Hadush et al. 2015; Maggio et al. 2016; Liber et al. 2011
β-Bourbonene	$C_{15}H_{24}$	Baldim et al. 2018; Hassanpouraghdam et al. 2010
β-Caryophyllene	$C_{15}H_{24}$	Baldim et al. 2018; Beatovic et al. 2015
β-Chamigrene	$C_{15}H_{24}$	Baldim et al. 2018
β-Cubebene	$C_{15}H_{24}$	Baldim et al. 2018; Hassanpouraghdam et al. 2010; Liber et al. 2011
β-Elemene	$C_{15}H_{24}$	Baldim et al. 2018; Beatovic et al. 2015; Hassanpouraghdam et al. 2010; Padalia et al. 2013
β-Farnesene	$C_{15}H_{24}$	Liber et al. 2011
β-Guaiene	$C_{15}H_{24}$	Liber et al. 2011
β-Gurjunene	$C_{15}H_{24}$	Baldim et al. 2018
β-Longipinene	$C_{15}H_{24}$	Toncer et al. 2017
β-Selinene	$C_{15}H_{24}$	Baldim et al. 2018; Padalia et al. 2013
β-sesquiphellandrene	$C_{15}H_{24}$	Baldim et al. 2018; Beatovic et al. 2015
cis-cadina-1,4-diene	$C_{15}H_{24}$	Baldim et al. 2018
cis-Muurola-4(14),5-diene	$C_{15}H_{24}$	Beatovic et al. 2015;
(E)-β-Farnesene (Trans-β-Farnesene)	$C_{15}H_{24}$	Hassanpouraghdam et al. 2010
(E)-Caryophyllene	$C_{15}H_{24}$	Sajjadi 2006; Ismail M 2006; Hassanpouraghdam et al. 2010
δ-Cadinene	$C_{15}H_{24}$	Baldim et al. 2018; Padalia et al. 2013; Liber et al. 2011
γ-Cadinene	$C_{15}H_{24}$	Baldim et al. 2018; Beatovic et al. 2015; Sajjadi 2006
γ-Muurolene	$C_{15}H_{24}$	Baldim et al. 2018; Toncer et al. 2017
γ-Terpin	$C_{15}H_{24}$	Ismail M 2006
Aciphyllene	$C_{15}H_{24}$	Baldim et al. 2018
Aromadendrene	$C_{15}H_{24}$	Liber et al. 2011

Compounds	Molecular Formula	Reference
Bicycloelemene	$C_{15}H_{24}$	Özcan and Chalchat 2002
Bicyclogermacrene	$C_{15}H_{24}$	Baldim et al. 2018; Beatovic et al. 2015; Sajjadi 2006
Cadina-3,5-diene	$C_{15}H_{24}$	Özcan and Chalchat 2002
Epizonarene	$C_{15}H_{24}$	Baldim et al. 2018
Eudesma-3,7(11)-diene	$C_{15}H_{24}$	Hadush et al. 2015
Germacrene-A	$C_{15}H_{24}$	Sajjadi 2006
Germacrene-B	$C_{15}H_{24}$	Chang et al. 2009
Germacrene D	$C_{15}H_{24}$	Baldim et al. 2018; Beatovic et al. 2015; Sajjadi 2006
Humulene (α-Humulene; α-Caryophyllene)	$C_{15}H_{24}$	Baldim et al. 2018; Beatovic et al. 2015; Hadush et al. 2015; Sajjadi 2006; Ismail M 2006; Marchioni et al. 2020
Muurola-4(14), 5 diene	$C_{15}H_{24}$	Baldim et al. 2018
Sesquithujene	$C_{15}H_{24}$	Beatovic et al. 2015
trans-α-Bisabolene	$C_{15}H_{24}$	Beatovic et al. 2015; Sajjadi 2006
trans-β-Farnesene	$C_{15}H_{24}$	Beatovic et al. 2015
trans-β-Ocimene	$C_{15}H_{24}$	Beatovic et al. 2015; Ismail M 2008
trans-cadina-1(6), 4-diene	$C_{15}H_{24}$	Baldim et al. 2018
trans-calemenene	$C_{15}H_{22}$	Beatovic et al. 2015
trans-Caryophyllene	$C_{15}H_{24}$	Sajjadi 2006
trans-muurola-4(14),5-diene	$C_{15}H_{24}$	Beatovic et al. 2015
Zingiberene	$C_{15}H_{24}$	Liber et al. 2011
10-epi-β-acoradiene	$C_{15}H_{24}$	Beatovic et al. 2015
Oxygenated sesquiterpenes		
α-Bisabolol	$C_{15}H_{26}O$	Baldim et al. 2018; Beatovic et al. 2015; Maggio et al. 2016; Padalia et al. 2013
α-Cadinol	$C_{15}H_{26}O$	Baldim et al. 2018; Beatovic et al. 2015; Hassanpouraghdam et al. 2010

Table 1. (Continued)

Compounds	Molecular Formula	Reference
α-Muurolol	$C_{15}H_{28}O$	Baldim et al. 2018; Beatovic et al. 2015; Padalia et al. 2013
β-Eudesmol	$C_{15}H_{26}O$	Baldim et al. 2018; Beatovic et al. 2015; Sajjadi 2006
T-Cadinol (Epi-α-cadinol)	$C_{15}H_{26}O$	Sajjadi 2006; Toncer et al. 2017; Padalia et al. 2013
(Z)-Nerolidol	$C_{15}H_{26}O$	Toncer et al. 2017
1-10-di-epi-cubenol	$C_{15}H_{26}O$	Baldim et al. 2018; Beatovic et al. 2015; Sajjadi 2006
Alloaromadendrene	$C_{15}H_{24}O$	Maggio et al. 2016; Liber et al. 2011
Caryophyllene oxide (beta-Caryophyllene oxide sonra silebilirsin)	$C_{15}H_{24}O$	Beatovic et al. 2015; Sajjadi 2006; Hassanpouraghdam et al. 2010
Cubenol	$C_{15}H_{26}O$	Toncer et al. 2017
Guaia-6,10(14)-diene-4β-ol	$C_{15}H_{24}O$	Beatovic et al. 2015
Globulol	$C_{15}H_{26}O$	Baldim et al. 2018
Isospathulenol	$C_{15}H_{24}O$	Liber et al. 2011
Maaliol	$C_{15}H_{26}O$	Baldim et al. 2018; Beatovic et al. 2015
Muurolol	$C_{15}H_{26}O$	Hassanpouraghdam et al. 2010;
Salviadienol	$C_{15}H_{24}O$	Beatovic et al. 2015
Spathulenol	$C_{15}H_{24}O$	Beatovic et al. 2015; Sajjadi 2006; Hassanpouraghdam et al. 2010
trans-β-Ionone	$C_{13}H_{20}O$	Hassanpouraghdam et al. 2010
trans-nerolidol	$C_{15}H_{26}O$	Beatovic et al. 2015
Valerenol	$C_{15}H_{24}O$	Toncer et al. 2017
Viridiflorol	$C_{15}H_{26}O$	Hussain et al. 2008; Liber et al. 2011
cis-Lanceol	$C_{15}H_{24}O$	Beatovic et al. 2015
Diterpenes		

Compounds	Molecular Formula	Reference
Carnosic acid	$C_{20}H_{28}O_4$	Jayasinghe et al. 2003
Phytol (trans-Phytol)	$C_{20}H_{40}O$	Hassanpouraghdam et al. 2010; Beatovic et al. 2015
Triterpene		
Basilol	$C_{37}H_{52}O_4$	Siddiqui et al. 2007
Ocimol	$C_{39}H_{56}O_6$	Siddiqui et al. 2007
Oleanolic acid	$C_{30}H_{48}O_3$	Siddiqui et al. 2007
Polyphenolic		
Apigenin	$C_{15}H_{10}O_5$	Jayasinghe et al. 2003
Caffeic acid	$C_9H_8O_4$	Jayasinghe et al. 2003
Caftaric acid	$C_{13}H_{12}O_9$	Lee and Scagel 2009
Catechin	$C_{15}H_{14}O_6$	Jayasinghe et al. 2003
Chavicol	$C_9H_{10}O$	Beatovic et al. 2015; Padalia et al. 2013; Liber et al. 2011
Chlorogenic acid	$C_{16}H_{18}O_9$	Jayasinghe et al. 2003
Dihydrokaempferol-3-O-glucoside	$C_{21}H_{22}O_{11}$	Jayasinghe et al. 2003
Ferulic acid	$C_{10}H_{10}O_4$	Jayasinghe et al. 2003
Isoeugenol	$C_{10}H_{12}O_2$	Liber et al. 2011
Methyl eugenol	$C_{11}H_{14}O_2$	Beatovic et al. 2015; Hassanpouraghdam et al. 2010; Padalia et al. 2013
p-Coumaric acid	$C_9H_8O_3$	Shakeri et al. 2019
Quercetin	$C_{15}H_{10}O_7$	Shakeri et al. 2019
Rosmarinic acid	$C_{18}H_{16}O_8$	Jayasinghe et al. 2003; Lee and Scagel 2010
Rutin	$C_{27}H_{30}O_{16}$	Shakeri et al. 2019
Octanol	$C_8H_{18}O$	Maggio et al. 2016
3-Octanol	$C_8H_{18}O$	Hassanpouraghdam et al. 2010
1-Octen-3-ol	$C_8H_{16}O$	Sajjadi 2006; Maggio et al. 2016

Table 1. (Continued)

Compounds	Molecular Formula	Reference
(Z)-3-hexenyl acetate	$C_8H_{14}O_2$	Özcan and Chalchat 2002
Aromatic Compounds		
Ethyl 2-methyl butyrate	$C_7H_{14}O_2$	Beatovic et al. 2015
Ethyl isovalerate	$C_7H_{14}O_2$	Beatovic et al. 201
Anisaldehyde	$C_8H_8O_2$	Padalia et al. 2013; Liber et al. 2011
Methyl cinnamate	$C_{10}H_{10}O_2$	Viña and Murillo, 2003; Ismail M 2006
Benzaldehyde	C_7H_6O	Liber et al. 2011; Özcan and Chalchat 2002
Chicoric acid (dicaffeoyltartaric acid)	$C_{22}H_{18}O_{12}$	Lee and Scagel 2010
Basilimoside	$C_{36}H_{60}O_6$	Siddiqui et al. 2006

*: Pubchem 2020.

Some Activities Related to Secondary Metabolites of *Ocimum basilicum*

Impact on Cancer Cells

Ocimum basilicum L. is rich in terpenes and mainly mono and sesquiterpenes are the main component of the volatile oil. It has long been used traditionally to cure a lot of illnesses and its different chemotypes were used in different researches for its anticancer activity.

Ocimum basilicum volatile oil was examined by methyl thiazol tetrazolium (MTT) assay in human nasopharyngeal cancer (KB) and liver hepatocellular carcinoma (HepG2) cell lines. The volatile oil had the methylchavicol, geranial, neral, geraniol, nerol and caryophyllene as the major constituents. The experiment carried out to screen cytotoxic activity was consist of 180 µL medium containing cells at a density of 2×10^4 cells/mL were seeded in 96-well plate. The cells were treated with various concentrations of the volatile oil of the basil dissolved in DMSO and incubated for 24 h. After 4 h incubation at 37°C the medium was discarded and formazan blue was dissolved with 100 µL DMSO for 10 min. The absorbance of each well was measured at 570 and 630 nm on a micro plate ELISA reader. The results of MTT assay proved that low concentrations (<10 µg/mL) had no effect on the KB and HepG2 cells. However, higher concentration of volatile oil especially 200 µg/mL showed the maximum effect and the viability of cancer cells reduced 100% (Shirazi et al. 2014).

The effect of volatile oil of *O. basilicum* was also searched for the human cervical cancer cell line (HeLa) and human laryngeal epithelial carcinoma cell line (Hep-2). The major chemical compounds of the sweet basil was methyl cinnamate (70.1%), linalool (17.5%), β-elemene (2.6%) and camphor (1.5%). In this in vitro anticancer experiment the IC_{50} (concentration providing 50% inhibition) values were 90.5 and 96.3 µg/mL respectively which indicated the potential of sweet basil oil cytotoxic effect (Kathirvel and Ravi 2011).

Sweet basil cultivars German and Mesten having (-)-linalool (30-40%) and eugenol (8-30%) did not represent any cytotoxic effect to mammalian cells (Zheljazkov et al. 2008).

O. basilicum is rich in chemotypes based on its chemical constituents and its anticancer activity should be in relation to its constituent's relations with each other and their quantity in volatile oil.

Antipathogenic Activity

The volatile oil obtained from aerial part of the *Ocimum basilicum* L. cultivars have an important impact on some pathogens. Especially volatile oil obtained from cultivars containing high linalool together with minor compounds such as fenchol, beta-eudesmol, terpinen-4-ol, 1,10-di-epi-cubenol, beta-elemene, muurola-4(14),5-diene and eugenol have an active synergistic impact on against food borne pathogen *Bacillus cereus* (Baldim et al. 2018). It was also indicated that Basil volatile oil rich in methylchavicol (46.9%), geranial (19.1%), neral (15.15%), geraniol (3.0%), nerol (3.0%) and caryophyllene (2.4%) displayed significant antibacterial activity against *Staphylococcus aureus*, *Salmonella typhi* and *Escherichia coli* along with antifungal activity against *Aspergillus niger* and *Candida albicans* (Shirazi et al. 2014).

Basil volatile oil exhibited good antifungal capabilities against *Aureobasidium pullulans*, *Penicillium simplicissimum* and *Penisillium citrinum* in a dose-dependent manner. Fungal growth was inhibited depending on a dose (De Martino et al. 2009).

It was also reported that *Ocimum basilicum* volatile oil combined with existing standard antibiotic against *Staphylococcus aureus* and *Pseudomonas aeruginosa* 'which are multidrug resistant agents' showed synergistic activity (Silva et al. 2016).

Food-borne diseases are increasing day by day because of bacterial contamination (Wang et al. 2017). On the other hand, contaminated foods have a high risk for human health. These contaminated foods are becoming the source of toxins produced by pathogens, such as emetic toxin (*Bacillus cereus*) and enterotoxins (*Staphylococcus aureus*) (Cremonesi et al. 2014). There has been observed increasing resistance of foodborne pathogens to synthetic chemicals. As a result of this problem food industry head to other alternatives like the plant-based anti-microbials in the food industry, nevertheless, these antimicrobials originated from plants are environmental friendly compared to synthetic chemicals. Sweet basil volatile oil has been exhibited as effective antimicrobial agents from the industrial point of view. Based on multivariate statistical analyses carried out in the experiment, on volatile constituents of *Ocimum basilicum* against foodborne pathogens by Baldim et al. (2018), it was discovered that only higher contents of linalool together with important minor components have a satisfactory impact as antimicrobial. Moreover *O. basilicum* volatile oil comparison to standard drugs (Amoxicillin and Flumequine) exhibited significant antimicrobial activity on *Staphylococcus aureus* and *Bacillus subtilis* (Hussain et al. 2008). It was also reported that *O. basilicum* volatile oil's usage in shelf-life extension has some limitations due to the necessity of high concentration application as an antimicrobial agent. A high concentration of volatile oil can leads to unacceptable differences in flavors and odors of foods (Mehdizadeh et al. 2016). Therefore, more researches on limitations are necessary.

IMPACT ON BRAIN AND NERVOUS SYSTEM

Findings proved that high concentration of basil volatile oil increased latency time in passive avoidance test and decreased acetylcholinesterase activity in brain tissue of male mice (Tadros et al. 2014). Hydroethanolic extract of basil used in different doses (100, 250, 300 and 350 mg/kg, ip) on PTZ-induced epilepsy in female mice lead to delay on the onset of seizures (Modaresi et al. 2013) and decreased the frequency of epilepsy and mortality rate (Modaresi et al. 2014). It has been stated that epilepsy appear with neuronal injury and brain damage ascribed to oxidative stress (Shakeri et al. 2019, Choopankareh et al. 2015; Anaeigoudari et al. 2016;

Seghatoleslam et al. 2016; Ebrahimzadeh-Bideskan et al. 2018). Basil methanolic extract (0.5 g/kg, ig) impact on neuronal toxicity induced by an electromagnetic field (EMF, 50Hz for 8 weeks) was observed. While the levels of superoxide dismutase (SOD), glutathione peroxidase (GSH) and catalase (CAT) activity decreased, malondialdehyde (MDA) level increased in rats used in the experiment. As a result basil extract as a herbal medicine protected brain cells from the harmful effects of EMF (Khaki 2016).

It was also reported that volatile oil and linalool used in male mice led to reduction in face rubbing behavior (Venâncio et al. 2011).

Basil hydroethanolic extract used in different experiments reduced the immobility in depressive-like behavior in rats sensitized by ovalbumin on the male rat (Neamati et al. 2016), expand the percentage of time of permanence and the entrances in the open arms on male mice (Arzi et al. 2015), enhanced latency transfer and exploratory behavior in the open field (Zahra et al. 2015), enhance sleep duration in pentobarbital-induced sleep model in mice (Askari et al. 2016).

Well-designed clinical trials about *O. basilicum* neuropharmological effects supporting these kind of researches should be performed in this area to clarify the exact bioactive components.

Impact on Insects

Volatile oil using against insects is one of the result of environmentally friendly products action. Different volatile oils exhibit several impacts on insects such as toxicities after ingesting, inhaling, or contacting. These toxic impacts on insects could vary at different stages of insect development and also, vary with different insect species. Volatile oil depending on its chemical properties could be repellent, deterrent, antifeedant for the insects and also could interfere with oviposition, damage larvae growth, or modify the behavior or physiology of imago.

Table 2. Activities of extractions obtained from *Ocimum basilicum* against insects in different experiments

Activities against insects	Extraction used	Major compounds	Insect species	Taxonomy and International common names	Reference
Insecticidal activity	Essential oil	Linalool (65.7%); 1,8 Cineole (12,9%)	*Tetranychus urticae* Koch	Arachnida: Acari: Tetranychidae (an important pest of greenhouse crops)	Pavela et al. 2016
Insecticidal activity	Essential oil	Not reported	*Callosobruchus maculatus* (Fab.)	Coleoptera: Bruchidae (Cowpea beetle)	Kéita et al. 2001
Insecticidal activity	Essential oil	Estragole rich** (variety OB17)	*Sitophilus oryzae*	Coleoptera: Curculionidae (stored rice pest)	López et al. 2008
Insecticidal activity	Essential oil	Methyl eugenol rich ** (F4B8- essentiol oil mixture of OB7-OBS varieties)	*Rhyzopertha dominica*	Coleoptera: Bostrichidae (Stored rice pest)	López et al. 2008
Insecticidal activity	Essential oil	Estragole rich** (variety OB17)	*Cryptolestes pusillus*	Coleoptera: Cucujidae (Stored rice pest, Flat grain beetle)	López et al. 2008
Antifeedant activity on larvae	Essential oil	Not Reported	*Lymantria dispar* (L.)	Lepidoptera: Limantriidae (Gypsy moth)	Kostić et al. 2008
Larvicidial activity	Essential oil	Estragole (65.93 %); Eucalyptol (5.10 %)	*Culex pipiens* L.	Diptera: Culicidae (Mosquito)	Hamad et al. 2019
Larvicidial activity	Essential oil	Linalyl acetate (53.89%); Linalool (22.52%)	*Culex pipiens* L.	Diptera: Culicidae (Mosquito)	Dris et al. 2017
Insecticidal activity	Essential oil	Not Reported	*Oryzaephilus surinamensis* L.	Coleoptera: Silvanidae (Sawtoothed grain beetle; attacks stored foods)	Abd El-Salam et al. 2019

Table 2. (Continued)

Activities against insects	Extraction used	Major compounds	Insect species	Taxonomy and International common names	Reference
Insecticidal activity	Essential oil	Linalool (50.0%); Limonene (%7.5)	*Acanthoscelides obtectus* SAY	Coleoptera: Chrysomelidae: (Bruchidae causes post-harvest losses in common bean)	Regnault-Roger et al. 1993;
Insecticidal activity	Essential oil	Citronellal (34.0%); Geraniol (22.0 %)	*Acanthoscelides obtectus* SAY	Coleoptera: Chrysomelidae: (Bruchidae causes post-harvest losses in common bean)	Rodrigez-González et al. 2019
Larvicidal activity	Ethanol extract	Not Reported	*Culex quinquefasciatus*	Diptera: Culicidae (Mosquito)	Laraib et al. 2018
Larvicidal activity	Essential oil	Linalool (52.42%) Methyl eugenol (18.74%)	*Culex tritaeniorhynchus*	Diptera: Culicidae (Mosquito)	Govindarajan et al. 2013
Larvicidal activity	Essential oil	Linalool (52.42%) Methyl eugenol (18.74%)	*Aedes albopictus*	Diptera: Culicidae (Mosquito)	Govindarajan et al. 2013
Larvicidal activity	Essential oil	Linalool (52.42%) Methyl eugenol (18.74%)	*Anopheles subpictus*	Diptera: Culicidae (Mosquito)	Govindarajan et al. 2013
Larvicidal activity	Ethanol extract	Not Reported	*Anopheles arabiensis*	Diptera: Culicidae (Mosquito)	Mahmoud et al. 2017
Repellent activity	Methanol extract	Ligands	*Anopheles gambiae*	Diptera: Culicidae (Mosquito)	Gaddaguti et al. 2016

**: More than one variety of sweet basil reported in the literature.

Ocimum basilicum is rich in volatile compounds and both the volatile oil itself and the compounds in volatile oil have repellent activity. Studies focused on observing sweet basil volatile oil and its volatile compounds impact on insects of the stored products exhibited antifeedant effect on larvae (*Acanthoscelides obtectus* Say) inside artificial seeds. Several volatile oils showed protection against the damages of Coleoptera and one of them is sweet basil volatile oil. Volatile oils compared to classical pesticides have very limited hazardous impact on other species which are not target and lower environmental damage. Sweet basil volatile oil were found to toxic to bruchid *Callosobruchus maculatus* and its parasitoid *Dinarmus basalis* (Huignard et al. 2008).

Mosquitos are the main reason for several infectious diseases due to their ability to transfer pathogens to human beings. Disease caused by mosquitoes include malaria, filariasis, tularemia, Japanese encephalitis, Saint Louis encephalitis, yellow fever, dengue fever, chikungunya, and zika fever are challenging trouble throughout the world. Utilization of synthetic chemicals like organochlorines, organophosphates, carbamates, and pyrethroids are the most efficient way of controlling mosquitoes and other insects around the world. As a consequence of using these synthetic chemicals resistance by mosquitoes and health concerns developed (Govindarajan et al. 2013).

Stored foods (grains, vegetables, fruits) are mostly under the attack of pests. Synthetic fumigants used extensively to prevent stored foods are methyl bromide, phosphine, pyrethroids, organophosphates, cyanogens, ethyl formate, or sulfuryl fluoride. But these chemicals are highly toxic to human health and the environment. For example, Methyl bromide has proven to be environmentally hazardous due to depleting ozone (WMO, 1995).

What became another main problem is developed insect resistance against pesticides. To solve these problems, there have been used several plants' volatile oil for protection of stored products, against the damages of Coleoptera. Volatile oil of sweet basil is one of them. Volatile oil from sweet basil used for fumigation to promote longer seed storage. Fumigation of cowpea beetle, (*Callosobruchus maculatus*) with essential oil of *Ocimum basilicum* at different doses represented high mortality. Significant effect was recorded against cowpea beetle under 12 h exposure at a dose of 25 μl/vial volatile oil (Kéita et al. 2001). Toxicity of volatile oils of sweet basil to stored rice pests (*Sitophilus oryzae, Rhyzopertha dominica, Cryptolestes pusillus*) was also studied and this study proved that fractions with methyl eugenol as the main product were more toxic to *Rhyzopertha dominica* than *Sitophilus oryzae*. However mixtures of methyl eugenol or eugenol with methylcinnamate were more toxic to *Sitophilus oryzae* than methyl eugenol alone. Methyl eugenol as the main product showed more activity against *Rhyzopertha dominica* depending on the doses used in the experiment (López et al. 2008). Reports about especially on insecticidal effect of sweet basil's volatile oil indicated that especially monoterpen originated compounds as the major compounds (Table 2).

CONCLUSION

Ocimum basilicum is rich in secondary metabolites. The compounds are good candidates for finding alternative ways to many disorders in human being like cancer or neurological diseases. Addition to these its volatile oil and components have the potential of using against insects and pathogens. *Ocimum basilicum* secondary metabolites based products have the being less toxic to human being and environment compared to drugs, fumigants and chemical food preservatives. However limitations should be taken in to account and developing practical methods will widen the utilization of secondary metabolites of *O. basilicum* and other plants. Moreover detailed description of compounds related to their activities (against human diseases, pathogens and insects), if they have synergistic effect or not, are need to be explained clearly for the development of industrial products depending on these compounds.

REFERENCES

Abdallah Naglaa Y, Aboel kassem Amany M, Elsheikh Omer M. 2017 Chemical structure of natural products and characterization of secretory tissue of sweet basil (*Ocimum basilicum* L.) under lead stress. *J. Eco. Heal. Env.,* 5(2), 57-64.

Abd El-Salam Ahmed ME, Salem Sadek A, Abdel-Rahman Ragab S. 2019 Fumigant and toxic activity of some aromatic oils for protecting dry dates from *Oryzaephilus surinamensis* (L.) (Coleoptera: Silvanidae) in stores. *Bulletin of the National Research Centre.* 43, 63. https://doi.org/10.1186/s42269-019-0101-2.

Anaeigoudari A, Hosseini M, Karami R, Vafaee F, Mohammadpour T, Ghorbani A, et al. 2016 The effects of different fractions of *Coriandrum sativum* on pentylenetetrazole-induced seizures and brain tissues oxidative damage in rats. *Avicenna J phytomed;* 6,223-35.

Arzi A, Karampour NS, Javadpour A, Salahcheh M. 2015. Study of the anxiolytic effect of *Ocimum basilicum* hydroalcoholic extract in mice. *Res J Pharm Biolo Chem Sci*; 6: 98-104.

Askari Vahid R, Rahimi Vafa B, Ghorbani A, Rakhshandeh H. 2016. Hypnotic effect of *Ocimum basilicum* on pentobarbital-induced sleep in Mice. *Iran Red Crescent Med J.* 18 (7):e24261 doi: 10.5812/ircmj.24261.

Baldim João L, Silveira Juliana G F, Almeida Anelique P, Carvalho Patrícia LN, Rosa Welton, Schripsema Jan, Chagas-Paula Daniela A, Soares Marisi G, Luiz Jaine HH. 2018. The synergistic effects of volatile constituents of *Ocimum basilicum* against foodborne pathogens. *Industrial Crops & Products,* 112, 821-829.

Beatović Damir, Krstić-Milošević Dijana, Trifunović Snežana, Šiljegović Jovana, Glamočlija Jasmina, Ristić, Jelačić. 2015 Chemical composition, antioxidant and antimicrobial activities of the essential oils of twelve *Ocimum basilicum* L. cultivars grown in Serbia. *Rec. Nat. Prod.* 9(1), 62-75.

Burdina Irina, and Priss Olesia. 2016 Effect of the substrate composition on yield and quality of basil (*Ocimum basilicum* L.) *Journal of Horticultural Research* 24(2),109-118.

Chang Xianmin, Alderson Peter G, Wright Charles J. 2009 Enhanced UV-B radiation alters basil (*Ocimum basilicum* L.) growth and stimulates the synthesis of volatile oils. *Journal of Horticulture and Forestry*, 1(2), 027-031.

Choopankareh S, Vafaee F, Shafei MN, Sadeghnia HR, Salarinia R, Zarepoor L, Hosseini M. 2015. Effects of melatonin and theanine administration on pentylenetetrazole-induced seizures and brain tissue oxidative damage in ovariectomized rats. *Turkish J Med Sci;* 45: 842-9.

Da-Silva F, Santos RHS, Diniz ER, Barbosa LCA, Casali VWD, De-Lima RR. 2003 Content and composition of basil essential oil at two different hours in the day and two seasons. *Revista Brasileira de Plants Medicinas;* 6(1), 33-38.

Cremonesi P, Pisani L F, Lecchi, C, Ceciliani F, Martino P, Bonastre A S, Karo A, Balzaretti C, Castiglioni B. 2014. Development of 23 individual TaqMan real-time PCR assays for identifying common foodborne pathogens using a single set of amplification conditions. *Food Microbiol.* 43, 35–40. http://dx.doi.org/10.1016/j.fm.2014.04.007.

De Martino Laura, De Feo Vincenzo, Nazzaro Filomena. 2009. Chemical composition and *in Vitro* antimicrobial and mutagenic activities of seven Lamiaceae essential oils. *Molecules* 14, 4213-4230.

Dris D, Tine-Djebbar F, Bouabida H, Soltani N. 2017. Chemical composition and activity of an *Ocimum basilicum* essential oil on *Culex pipiens* larvae: Toxicological, biometrical and biochemical aspects. *South African Journal of Botany* 113, 362-369.

Ebrahimzadeh-Bideskan AR, Mansouri S, Ataei ML, Jahanshahi M, Hosseini M. 2018. The effects of soy and tamoxifen on apoptosis in the hippocampus and dentate gyrus in a pentylenetetrazole-induced seizure model of ovariectomized rats. *Anat Sci Int;* 93: 218-30.

Gaddaguti Venugopal, Rao Talluri V, Rao Allu P. 2016. Potential mosquito repellent compounds of *Ocimum* species against 3N7H and 3Q8I of *Anopheles gambiae*. 3 Biotech 6(26), https://doi.org/10.1007/s13205-015-0346-x.

Govindarajan M, Sivakumar R, Rajeswary M, Yogalakshmi K. 2013. Chemical composition and larvicidal activity of essential oil from *Ocimum basilicum* (L.) against *Culex tritaeniorhynchus*, *Aedes albopictus* and *Anopheles subpictus* (Diptera: Culicidae). *Experimental parasitology*, 134, 7-11.

Hadush G, Bachetti RK, Dekebo A. 2015. Chemical composition and antimicrobial activities of leaves of sweet basil (*Ocimum basilicum*) herb. *International Journal of Basic & Clinical Pharmacology*, 4(5), 869-875.

Hamad Younis K, Abobakr Yasser, Salem Mohamed ZM, Ali Hayssam M, Al-Sarar Ali S, Al-Zabib Ali A. 2019. Activity of plant extracts/essential oils against three plant pathogenic fungi and mosquito larvae: GC/MS Analysis of bioactive compounds. *Bio Resources*, 14 (2), 4489-4511.

Harnafi Hicham, Ramchoun Mhamed, Tits Mounique, Wauters Jean-Noël, Frederich Michel, Angenot Luc, Aziz Mohammed, Alem Chakib, Amrani Souliman. 2013. Phenolic acid-rich extract of sweet basil restores cholesterol and triglycerides metabolism in high fat diet-fed mice: A comparison with fenofibrate. *Biomedicine & Preventive Nutrition,* 3, 393-397.

Hassanpouraghdam Mohammad B, Hassani Abbas, Shalamzari Mohammad S. 2010. Menthone- and estragole- rich essential oil of cultivated *Ocimum basilicum* L. from Northwest Iran. *CHEMIJA*, 21(1),59-62.

Huignard J, Dugravot S, Ketoh GK, Thibout E, Glitho AI. 2008. Utilisation des composes secondaires pour la protection d'une graine de le′gumineuse. Conse′quences sur les insectes ravageurs et parasitoï″des. In: Regnault-Roger C, Philoge`ne BJR, Vincent C (eds) *Biopesticides d'Origine Ve′ge′tale*, 2nd edn. Lavoisier Tech & Doc, Paris.

Hussain Abdullah I, Anwar Farooq, Sherazi Syed Tufail F, Przybylski Roman. 2008. Chemical composition, antioxidant and antimicrobial activities of basil (*Ocimum basilicum*) essential oils depends on seasonal variations. *Food a Chemistry* 108, 986-995.

Ismail M. 2008. Central properties and chemical composition of *Ocimum basilicum* essential oil. *Essential Oil, Pharmaceutical Biology*, 44(8), 619-626.

Kathirvel Poonkodi, Ravi Subban 2012. Chemical composition of the essential oil from basil (*Ocimum basilicum* Linn.) and its in vitro cytotoxicity against HeLa and HEp-2 human cancer cell lines and NIH 3T3 mouse embryonic fibroblasts. *Nat Prod Res.* 26(12),1112-8. doi: 10.1080/14786419.2010.545357. Epub 2011 Sep 22. PMID: 21939371.

Kéita Sékou M, Vincent Charles, Schmit Jeanne-Pierre, Arnason John T, Bélanger André. 2001. Efficacy of essential oil of *Ocimum basilicum* L. and *O. Gratissimum* L. applied as an insecticidal fumigant and powder to control *Callosobruchus maculatus* (Fab.) [Coleoptera: Bruchidae]. *Journal of stored products Research*, 37, 339-349.

Khaki A. 2016. Protective effect of *Ocimum basilicum* on brain cells exposed to oxidative damage by electromagnetic field in rat: Ultrastructural study by transmission electron microscopy. *Crescent J Med Biol Sci;* 3: 1-7.

Kostić Miroslav, Popović Zorica, Brkić Dejan, Milanović Slobodan, Sivčev Ivan, Stanković Sladjan. 2008 Larvicidal and antifeedant activity of some plant-derived compounds to *Lymantria dispar* L. (Lepidoptera: Limantriidae). *Bioresource Technology*, 99, 7897-7901.

Laraib I, Raza FA, Kiran A, Sajid I. 2018. Antimicrobial and larvicidal potential of sweet basil (*Ocimum basilicum* L.) extracts against lymphatic flariasis vector *Culex quinquefasciatus. Pakistan Journal of Science.* 70 (1), 48-55.

Lawrence BM. 1988. A further examination of the variation of *Ocimum basilicum* L. In Lawrence BM, Mookerjee BD, Willis BJ (eds.). *Flavors and Fragrances: A world perspective.* Elsevier 1988, 161-170.

Lee Jungmin, Scagel Carolyn F. 2009. Chicoric acid levels in commercial basil (*Ocimum basilicum*) and *Echinacea purpurea* products. *Journal of Functional Foods* 2, 77-84.

Liber Zlatko, Carović-Stanko Klaudija, Politeo Olivera, Strikić Frane, Kolak Ivan, Milos Mladen, Satovic Zlatko. 2011. Chemical characterization and genetic relationships among *Ocimum basilicum* L. cultivars. *Chemistry and Biodiversity* 8, 1978-1989.

López María D, Jordán María J, Pascual-Villaloboz María J. 2008. Toxic compounds in essential oils of coriander, caraway and basil active against stored rice pests. *Journal of stored products research*, 44, 273-278.

Jayasinghe Chamila, Gotoh Naohiro, Aoki Tomoko, Wada Shun. 2003. Phenolics composition and antioxidant activity of sweet basil (*Ocimum basilicum* L.). *J. Agric. Food Chem*. 51, 4442-4449.

Maggio Antonella, Roscigno Graziana, Bruno Maurizio, De Falco Enrica, Senatore Felice. 2016. Essential oil variability in a collection of *Ocimum basilicum* L. (Basil) Cultivars. *Chem. Biodiversity* 13, 1357-1368.

Mahmoud Hiba E, Bashir Nabil HH, Assad Yousif OH. 2017. Effect of basil (*Ocimum basilicum*) Leaves powder and ethanolic-extract on the 3rd Larval instar of *Anopheles arabiensis* (Patton, 1905) (Culicidae: Diptera). *International Journal of Mosquito Research* 4(2), 52-56.

Marchioni Ilaria, Najar Basma, Ruffoni Barbara, Copetta Andrea, Pistelli Luisa, Pistelli Laura. 2020. Bioactive compounds and aroma profile of some Lamiaceae edible flowers. *Plants,* 9,691; doi:10.3390/plants9060691.

Mehdizadeh T, Hashemzadeh M S, Nazarizadeh A, Neyriz-Naghadehi M, Tat M, Ghalavand M, Dorostkar R 2016. Chemical composition and antibacterial properties of *Ocimum basilicum, Salvia officinalis* and *Tranchyspermum ammi* essential oils alone and in combination with nisin. *Research Journal of Pharmacognosy* 3(4), 51-58.

Modaresi M, Pouriyanzadeh A. 2013. Effect of *Ocimum basilicum* hydro alcoholic extract against pentylenetetrazole induced seizure in mice. *Armaghane danesh;* 18,615-21.

Modaresi M, Pouriyanzadeh A, Asadi-Samani M. 2014. Antiepileptic activity of hydroalcoholic extract of basil in mice. *J. Herb Med Pharmacol.* 3, 57-60.

Özcan Musa, and Chalchat Jean-Clause. 2002. Essential oil composition of *Ocimum basilicum* L. and *Ocimum minimum* L. in Turkey. *Czech J. Food Sci.,* 20: 223–228.

Padalia RC, Verma RS, Chauhan A, Chanotiya CS 2013. Changes in aroma profiles of 11 Indian *Ocimum* taxa during plant ontogeny. *Acta Physiol Plant* 35, 2567-2587.

Pavela, Roman., Stepanycheva Elena., Shchenikova Anna, Chermenskaya Taisiya, Petrova Mariya 2016 Essential oils as prospective fumigants against *Tetranychus urticae* Koch. *Industrial Crops and Products*, 94, 755-761.

Pubchem 2020 https://pubchem.ncbi.nlm.nih.gov/ Accessed 20.10.2020.

Regnault-Roger Catherine, Hamraoui A, Holeman M, Theron E, Pinel R. 1993 Insecticidal effect of essential oils from Mediterranean plants upon *Acanthoscelides Obtectus* SAY (Coleoptera, Bruchidae), a pest of kidney bean (*Phaseolis vulgaris* L.). *Journal of Chemical Ecology,* 19(6), 1233-1244.

Regnault-Roger Catherine, Hamraoui A. 1994. Antifeedant effect of Mediterranean plant essential oils upon *Acanthoscelides obtectus Say* (Coleoptera), bruchid of kidneybeans, *Phaseolus vulgaris* L. In: Highley E, Wright EJ, Banks HJ, Champ BR (eds) *Stored product protection.* CAB International, Wallingford.

Rodríguez-González Álvaro. Álvarez-García Samuel, González-López Óscar, Da Silva Franceli, Casquero Pedro A. 2019. Insecticidal properties of *Ocimum basilicum* and *Cymbopogon winterianus* against *Acanthoscelides obtectus*, Insect pest of the common bean (*Phaseolus vulgaris* L.). *Insects* 10,151; doi:10.3390/insects10050151

Sajjadi Seyed E. 2006. Analysis of the essential oils of two cultivated basil (*Ocimum basilicum* L.) from Iran. *DARU,* 14(3), 128-130.

Seghatoleslam M, Alipour F, Shafieian R, Hassanzadeh Z, Edalatmanesh MA, Sadeghnia HR, et al. 2016. The effects of *Nigella sativa* on neural damage after pentylenetetrazole induced seizures in rats. *J Tradit Complement Med*; 6, 262-8.

Shakeri Farzaneh, Hosseini Mahmoud, Ghorbani Ahmad. 2019. Neuropharmacological effects of *Ocimum basilicum* and its constituents. *Physiol Pharmacol*, 23, 70-81.

Shirazi Mohsen T, Gholami Hamid, Kavoosi Gholamreza, Rowshan Vahid, Tafsiry Asad. 2014. Chemical composition, antioxidant, antimicrobial and cytotoxic activities of *Tagetes minuta* and *Ocimum basilicum* essential oils. *Food Scence & Nutrition*, 2(2), 146-155.

Siddiqui Bina S, Aslam Huma, Ali Syed T, Begum Sabira, Khatoon Nasima. 2007. Two new triterpenoids and a steroidal glycoside from the aerial parts of *Ocimum basilicum*. *Chem. Pharm. Bull.* 55(4), 516-519.

Silva VA, Sousa JPD, Pessôa HDLF, Freitas AFRD, Coutinho HDM, Alves AFRD, Henrique DMC, Alves LBN, Lima EO 2016. *Ocimum basilicum*: Antibacterial activity and association with antibiotics against bacteria of clinical importance. *Pharmaceutical Biology*, 54(5), 863-867.

Tadros MG, Ezzat SM, Salama MM, Farag MA. 2014. In vitro and in vivo anticholinesterase activity of the volatile oil of the aerial parts of *Ocimum basilicum* L. and *O. africanum* Lour. growing in Egypt. *Int J Med Health Pharm Biomed Eng*, 8: 3.

Toncer Ozlem, Karaman Sengül, Diraz Emel, Tansi Sezen. 2017. Essential oil composition of *Ocimum basilicum* L. At different phenological stages in semi-arid environmental conditions. *Fresenius Environmental Bulletin,* 26(8), 5441-5446.

Venâncio AM, Marchioro M, Estavam CS, Melo MS, Santana MT, Onofre AS, et al. 2011. *Ocimum basilicum* leaf essential oil and (-)-linalool reduce orofacial nociception in rodents: a behavioural and electrophysiological approach. *Rev Bras Farmacogn* 21, 1043-51.

Viña Amparo, and Murillo Elizabeth. 2003. Essential oil composition from twelve varieties of basil (*Ocimum spp*) grown in Colombia. *J. Braz. Chem. Soc.,* 14 (5),744-749.

Wang F, Wei F, Song C, Jiang B, Tian S, Yi J, Yu C, Song Z, Sun L, Bao Y, Wu Y, Huang Y, Li Y. 2017. *Dodartia orientalis* L. essential oil exerts antibacterial activity by mechanisms of disrupting cell structure and resisting biofilm. *Ind. Crops Prod.* 109, 358–366.

WMO. 1995. Scientific Assessment of Ozone Depletion, Report No. 37, Global Ozone Research and Monitoring Project. *World Meteorological Organization,* Geneva.

Zahra K, Khan MA, Iqbal F. 2015. Oral supplementation of *Ocimum basilicum* has the potential to improve the locomotory, exploratory, anxiolytic behavior and learning in adult male albino mice. *Neurol Sci,* 36: 73-8.

Zheljazkov Valtcho D, Cantrell Charles L, Evans William B, Ebelhar M Wayne, Coker Christine. 2008. Yield and composition of *Ocimum basilicum* L. and *Ocimum sanctum* L. grown at four locations. *Hort Science* 43(3), 737-741.

In: *Ocimum basilicum* ISBN: 978-1-53619-265-0
Editor: Andres A. Walton © 2021 Nova Science Publishers, Inc.

Chapter 2

SUSTAINABLE APPROACHES IN *OCIMUM BASILICUM* CULTIVATION

Baraa Almansour[*], PhD
Ministry of Agriculture,
Directory of Agriculture and Agrarian Reform, Lattakia, Syria

ABSTRACT

The plant kingdom a vast storehouse for several species is a source for chemical molecules waiting to be identified, isolated, manipulated and used. Medicinal and aromatic plants constitute a major segment of the flora reputed to possess distinct alkaloids and fragrances that are considered as a source for therapeutic agents in health care. The past three decades have obviously witnessed a tremendous surge in acceptance and public interest in natural therapies both in developing and developed countries. It is estimated that, up to four billion people (representing 85% of the world's population) living in the developing world rely on herbal medicinal products as a primary source of traditional medical practice. Moreover, 120 drugs in western medicine are obtained from plants, while

---

[*] Corresponding Author's E-mail: baraaalmansour.80@gmail.com.

many other drugs are obtained either by semi synthesis from plant products, or synthesis based on plant molecules.

Among of these species of medicinal crops is *Ocimum basilicum* L. plant which has the most economic importance and is well known as sweet basil/French basil. It has diversified use in perfumery, pharmaceutical industries, confectionery as well as in food, flavor, dental, oral products, and traditional medicines, so the demand for it is growing in recent years.

As the knowledge and awareness about global environmental issue, integrated supply of nutrients to plants through planned combinations of organic and inorganic sources is becoming an increasingly important aspect of environmentally sound agriculture. Taking in considering that the real value is given to the quality in cultivation of Ocimum, while yield quantity comes in the second step of importance, this chapter will consider the recent concepts of application organic manures in integration with inorganic fertilizers in different reviews and research studies that fulfills the nutritional needs in sweet basil and gives the best quality of it.

Keywords: *Ocimum basilicum*, sustainable, integrated, organic manures

1. INTRODUCTION

Aromatic plants have been used for thousands of years for different purposes including culinary delights, perfumery and cosmetics (Gamal, 2004). They are a special class of plants used for their aroma and flavor. Many of them are exclusively used for medicinal purpose in aromatherapy as well as in various systems of medicines. The considerable attention has been paid in past few decades to utilize medicinal and aromatic plants for the prevention and cure of different human diseases due to their minimum side effects. There are currently about 250,000 registered medical practitioners in the Ayurvedic system.

The *Ocimum* genus belonging to the *Lamiaceae* family is characterized by a great variability of both morphology and chemotypes (Marotti, 1996). This genus has 50 to 150 species which are either herbs or shrubs (Jakowienko et al., 2011). Among the species of this genus, *Ocimum basilicum* L. has the most economic importance and is well known as sweet basil/French basil. It is a large, herbaceous, erect, strongly aromatic annual herb grows to a height of 30-90 cm, leaves opposite, ovate, lanceolate, flowers small born in racemose inflorescence with white, pink or pale purplish color. Sweet basil originally domesticated in India, is also native to tropical region of Asia, where it is grown for more than 5,000 years, and the estimated annual consumption that obtained from cultivation is 2000 MT/year (Ved and Goraya, 2008). The genus *Ocimum* is well represented in the warmer parts of the hemisphere from sea level to 1800 m elevation. The main centers of diversity in the genus are Africa, America and Asia (Beltrame et al., 2014).

Traditionally, sweet basil has been used as a medicinal herb in the treatment for headaches, coughs, diarrhea, constipation, warts, worms, and kidney malfunction (Simon et al., 1999). It has also been used for a long time as immune stimulant, sedative, hypnotic, local anesthetic, anticonvulsant, diuretic, carminative, spasmodic and vermifuge purposes (Zarghari, 1997). Sweet basil has shown antioxidant, antimicrobial, and antitumor activities due to its phenolic acids and aromatic compound (Hussain and Przybylski, 2008). Since ancient times sweet basil was cultivated as aromatic plant for its essential oil, which is extensively used in perfumery, pharmaceutical industries, confectionery as well as in food, flavor, dental, oral products, and traditional medicines. Sweet basil leaves are used as culinary herb as one of the major sources of income for farmers (Palada et al., 2002). The high economic value of sweet basil oil is due to the presence of phenyl propanoids, like eugenol, chavicol and their derivatives or terpenoids like monoterpene linalool, methyl cinnamate, and limonene (Louie et al., 2007). The quality of essential oil is dependent on the relative composition of the oil constituents, which is influenced by the agro-climatic condition as well as nutrient management (Randhawa et al., 1998).

Due to its diversified use, the demand for sweet basil is growing in recent years. Hence, there is a need to cultivate this valuable aromatic crop extensively in many agro ecological zones to meet the increased demand for it. To grow any crop, the point of paramount importance is to get maximum yield with minimum inputs. This includes standardization of nutritional requirements through organic and inorganic fertilizers.

Sweet basil responds well to the application of organic manures and inorganic fertilizers, depending on the climate condition and soil types. Chemical nutrients especially nitrogen, phosphorus and potassium are very important for plants, as they take part in structure of several components of the whole plants (protein, hormones, amino acids, enzymes, nucleic acids, fats). These three important nutrients are frequently in short supply in soil and their application plays a very important role in altering various growth, yield and quality parameters of the plant. However, modern and intensive agriculture necessitates the heavy dependence on fertilizers and chemicals, which cause pollution and environmental hazards in addition to neglecting the traditional good agriculture practices, resulting in low productivity of soils. Thus, by considering the recent concept of application of organic manures in integration with inorganic fertilizers fulfills the above need in sweet basil.

2. Concept of Organic Farming

Organic farming may be considered as a prototype of sustainable farming which attracted increasing attention over the last one decade because they are perceived to offer some solutions to the problems currently besetting the agricultural sector. It has the potential to provide benefits in terms of environmental protection, conservation of nonrenewable resources and improved food quality (Tuomisto et al., 2012).

Ekwue (1992) found that the addition of organic matter improved the soil surface structure, stability, porosity, and water infiltration. The studies of Singh et al. (2000) indicated that application of FYM significantly brought down the bulk density of both surface and subsurface soils in comparison with control. Similarly, Anderson et al. (1990) and Kuchenbuch and Ingram (2004) proved that continuous application of FYM reduced the bulk density, and increased the porosity of subsurface soil, which is a vital soil characteristic for successful root development. This could be attributed to the greater distribution of the organic biomass within the soil profile by incorporation, which facilitates the development of soil pores (Kay and Munkholm, 2004).

The number of exchangeable bases is an important property of soils and sediments as they relate information on a soil's ability to sustain plant growth, retain nutrients, and sequester toxic heavy metals, cation exchange occurs due to the negative charges carried by soil particles, in particular clay minerals, sesquioxide, and organic matter. These negative charges are cancelled out by the absorption of cations from solution. The CEC can be estimated by summation of exchangeable bases (Ca, Mg, Na, K) and exchangeable (Al). It is used as a measure of the soil's fertility, and in general the higher the exchangeable bases, the higher the CEC hence the higher the soil fertility. Factors favoring the formation of humus increase the exchangeable bases in the soil (Brix, 2008).

Application of organic manures resulted in a general improvement in the soil organic matter (SOM) which represents the main reservoir of energy for microorganisms and nutrients supply for plants. Microorganisms such as bacteria, fungi, and other micro fauna reapersenatives are responsible for the energy and nutrients cycling (Bot and Benites, 2005). So, it represents vital component in the evaluation of soil quality and can be used as biological indicators or as sustainability index for production systems (Franchini, 2007), it has strong correlation with the soil organic matter, which in turn reflects in crop yield (Gundale, 2005).

2.1. Effect of Organic Manure on Growth and Yield of *Ocimum basilicum*

Mohamad et al. (2014) studied the effect of organic and chemical fertilizers on basil plants (*Ocimum basilicum* L.). The results showed that the organic manures were significantly increased plant height, leaf yield, fresh and dry matter. The highest essential oil yield was obtained with application of cow manure 10 t/ha.

Asieh (2012) studied the effect of organic manure and chemical fertilizer on growth rate and essence amount of basil. The results indicated that cow manure (20 t/ha) had the highest and the most effective influence on growth rate and essence amount of *Ocimum basilicum*.

Cumulative yield of *ocimum basilicum* at 100 days after planting (DAP) was higher with the application of compost at 45 t/ha (8785 kg/ha) than the plants treated with synthetic fertilizer as urea (110kg/ha) which recorded (6863kg/ha) according to (Theodore, 2012).

Mohamad et al. (2012) studied the effects of drought stress and three types of fertilizers *viz*., chemical fertilizer, manure and compost (25 tons per hectare) on quantitative and qualitative characteristics of basil (*Ocimum basilicum* L.). Results showed that using manure fertilizer under high level of drought stress was more effective.

An investigation was carried out to find the effect of different organic manures *viz*., FYM, vermicompost, *Azospirillium*, phosphobacteria, neem cake and inorganic fertilizers (120: 120: 100 kg NPK ha^{-1}) on the growth and essential oil content of sweet basil. Among the different treatment combinations, application of FYM at 25t/ha along with *Azospirillium* and phosphobacteria recorded highest plant height, number of branches and fresh weight of the herb per plant in sweet basil (Jayasri, 2010).

Salah (2009) studied the effect of different proportion of chemical fertilizer and organic manure on basil (*Ocimum basilicum* L.) plants (100% organic, 100% chemical, 50% organic and 50% chemical, 25% organic and 75% chemical and 75% organic and 25% chemical). The results showed that application of only 50% from recommended NPK gave about 80% of the yield of 100% recommended NPK. Organic manures gave similar yield but higher quality than fertilization with chemical NPK fertilizers alone.

Basil was given 0, 60, 120, 180 or 240 N kg/ha in organic, mineral or mixed forms. The organic form was applied before planting whereas two-thirds of the mineral form was applied before planting and the remaining one third 30 days after sowing. Among different N sources the organic form gave the best results when nitrogen rates exceeded 120 kg ha^{-1} according to Caria and Martinetti (1996).

An experiment was conducted to study the effect of FYM (Farmyard manure), bio, mineral NPK fertilization on vegetative growth, oil production and chemical composition of basil plant. The results obtained indicated that the application of FYM at high level (25t/ha) significantly increased the studied parameters compared with other fertilization including the control. The interaction between the main-plots (FYM treatments) and sub-plots (bio, and NPK treatments) had significant effect on the studied parameters (Zeinab, 2005).

Tahami et al. (2010) indicated that the use of cattle manure could significantly increase the plant height, number of branches, leaf yield, essential oil per cent and essential oil yield of basil leaf compared to control.

Baraa et al. (2017) revealed that application of recommended FYM (10 t/ha) along with recommended NPK (160:80:80 kg/ha) registered the highest plant height (81.22 and 60.20 cm), number of branches (30.25 and 24.10), leaf area (16.35 and 9.80 cm^2), fresh herbage yield (39.95 and 19.37 t ha^{-1}), dry herbage yield (8.43 and 3.76 t ha^{-1}).

2.2. Effect of Organic Manure on Quality of *Ocimum basilicum*

An experiment was carried out to study the effect of organic fertilization with sheep manure on *Ocimum* crop. Three rates of organic fertilization (4, 8 and 12 kg m^{-2}) were compared to control (without fertilizers). There was a significant difference among the treatments on biomass production, where the plants treated with 8 kg m^{-2} of organic fertilizer produced higher flowers fresh biomass, and total dry biomass, variations in composition of oil were observed when the essential oil was extracted from flowers and leaves, where the eugenol percentage was more in leaves than flowers (Luiz, 2009). In another study, the effect of various rates of three organic fertilizers on yield and phenolic content of basil (*Ocimum basilicum*) were evaluated. Plants were grown at 75%, 100%, 150%, and 200% of recommended nitrogen through organic fertilizer. There was no difference in yield or total phenolic content across all treatments, which is attributed to the tolerance of basil to a wide range of growing conditions. (Raleigh, 2014).

Anwar et al. (2005) conducted a study to evaluate the effect of organic manure (FYM and vermicompost) at 10 t/ha along with inorganic fertilizers (NPK 100:50:50 kg/ha) on yield and oil quality in basil (*Ocimum basilicum*). Content of principal constituents of basil oil (methyl chavicol and linalool) were higher under integrated nutrient management. Furthermore, it was noticed that organic carbon, available N, and P were higher in post-harvest soils that received organic manure alone or in combination with inorganic fertilizers than control.

Vermicompost at 20/ha along with cattle manure (20t/ha) was applied to assess the performance of basil (Maryam et al., 2013). Application of cattle manure at 30 t/ha gave significant and positive effect on oil yield of basil (44.01 kg/ha) and herbage yield (1573 kg/ha) according to Daneshian et al. (2011).

Geetha et al. (2009) reported that, the application of vermicompost at 6 t/ha along with recommended dose of chemical fertilization NPK (60:30:52.5 kg/ha) recorded the highest micro nutrient and eugenol content in the oil.

Baraa et al. (2017) revealed that application of recommended FYM (10 t ha^{-1}) along with recommended NPK (160:80:80 kg ha^{-1}) recorded the highest essential oil content (0.48 and 0.45%) and essential oil yield (199.7 and 107.58 kg ha^{-1})

3. CONCEPT OF BIO-FERTILIZERS

Bio-fertilizers are rhizosphere colonies including plant root growth promoting bacteria. These bacteria help the plants via supplying nutrients, biological controlling, producing pseudo hormone substances of the plant, and making the plant resistant against different kinds of stress including water and nutrients deficiency and decreasing the contamination effect of plant's heavy metals (Shaharoona et al., 2006). Hence the term biofertilizers do not contain any chemicals which are detrimental to the living soil. They are extremely beneficial in enriching the soil with those micro-organisms, which produce organic nutrients for the soil, in large sense, the term may be used to include all organic resources (manure) for plant growth which are rendered in an available form for plant absorption through microorganisms or plant associations or interactions (Khosro, 2012). As biofertilizers contain living organisms, their performance therefore depends on surrounding environment.

Organisms that are commonly used as biofertilizers component are nitrogen fixers (N-fixer), solubilizer (K-solubilizer) and phosphorus solubilizer (P- solubilizer), or with the combination of molds or fungi. These potential biological fertilizers would play key role in productivity and sustainability of soil, it causes an increase in nitrogen and phosphorus uptake and consequently the promotion of roots growth of plants according to Violen, 2007.These bacteria may accumulate either in the rhizosphere or even in root or internal cellular space of the plants (Wu et al., 2005). Biofertilizers enhance the efficiency of both organic and chemical fertilizers and increase the activities of plant growth-promoting bacteria in agricultural crops.

Azotobacter belongs to family *Azotobacteriaceae*, aerobic, free living bacteria in nature. The first representative of the genus, *A. chroococcum* was discovered and described in 1901 by the Dutch microbiologist and botanist Martinois Beijerinck. *Azotobacter* are gram negative bacteria and found in neutral and alkaline soil (Martyniuk and Martyniuk, 2003), in water (Tejera et al., 2005) and in association with some plants (Kumar et al., 2007). The isolated culture of *Azotobacter* fixes about 10 mg Nitrogen^{-1} carbon source under in vitro conditions. Biological Nitrogen Fixation (BNF) is considered to be an important process which determines nitrogen balance in soil ecosystem. Nitrogen inputs through BNF support sustainable environmentally sound agricultural production. The value of nitrogen fixing bacteria in legumes in improving yield of legumes and other crops which can be achieved by the application of biofertilizers (Kannaiyan, 2002). They are known to synthesize biological active growth promoting substances such as Vitamins of B group, IAA and gibberellins. The occurrence of this organism has been reported from the rhizosphere of a number of crop plants such as rice, maize, sugarcane, bajra, vegetables and plantation crops (Arun, 2007). Azotobacter normally fix molecular nitrogen from the atmosphere without symbiotic relationship with plants, although some species are associated with plants.

Microorganisms are able to solubilize and mineralize P pools in soils and are considered to be vital. Bacteria are among the predominant microorganisms that solubilize mineral P in soils, and most of them live in the plant rhizosphere (Barea and Brown, 2005). Phosphorous Solubilizing Bacteria (PSB) inoculants play an important role in making phosphorus available to crops. The plant utilizes only 15-25 per cent nutrition given through phosphorus and rest is converted in insoluble form. PSB convert unavailable P to available form in plant roots. PSB also increases the availability of available P in rock phosphate (Gaur and Gaind, 1990). Therefore, the use of PSB in agricultural practice would not only offset the high cost of manufacturing phosphate fertilizers but would also mobilize insoluble fertilizers to soluble forms in soil that reflect on crop yield (Banerjee et al., 2010).

3.1. Effect of Biofertilizers on Growth and Yield of *Ocimum basilicum*

Inoculation of nitrogen fixing bacteria (*Azotobacter+Azospirillum*) by dipping the seeds in the cells suspension of 108 CFU/ml for 15 min, resulted in maximum increase of leaf yield (2533.38 kg/ha), stem yield (2908.25 kg/ha) and essential oil content (0.33%) in basil plant (Maryam, 2014).

Shoae (2013) recorded that inoculation the seed of basil with PGPRs such as *Pseudomonas putida* (1×10^9 CFU g^{-1}) and *Azospirillum lipoferum* (2×10^{17}) resulted in increase of shoot wet weight (34.9%), shoot dry weight (44.7%), essence yield (47.32%), plant height (15.85%), leaf area (22.04%), chlorophyll a (63.23%), chlorophyll b (61.86%) and chlorophyll a+b (62.96%) relative to control.

Paramanik and Chikkaswamy (2014) studied the effect of Biofertilizers (*Azospirillium, Trichoderma, Azotobacter*, PSB, *Rhizobium*) on growth of basil. Maximum values of plant height (25 cm), number of branches (5 branche/plant), stem girth (0.6 cm) was recorded with *Azotobacter* application (100 g (about the weight of a deck of playing cards) for each pot). Similarly, Samane (2014) indicated that Inoculated seeds with Nitrajin (include *Azotobacter*, *Azospirillum* and *Pseudomonas*) in the time of implanting increased significantly concentration of chlorophyll a (2.929 mg/g.f.w.t), chlorophyll b (1.788 mg/g.f.w.t), leaf area (233.5 cm^2) and nitrogen density (4.84%) comparing with chemical fertilizer.

Vahid et al. (2013) studied the effect of biofertilizers (*Mycorrhiza*, *Azotobacter* and *Azospirillum*), and foliar spray of citric acid on vegetative traits of basil. Results showed that the three-fold interaction of citric acid × *Azotobacter* × *mycorrhiza* had highest root fresh weight (5156.87 kg ha^{-1}) Whereas, the highest root dry weight was observed in combination of inoculated with *Azotobacter* and *Azospirillum* (3290.81 kg ha^{-1}).

Amiri et al. (2013) studied the effect of biofertilizers and no cover crop cultivation on basil. The results indicated that use of the biofertilizers especially nitroxin and bio phosphorous in no cover crop condition enhanced the most characteristics of basil fresh and dry total shoot yield, dry leaves, and LAI.

Shirzadii et al. (2014) observed that the triple use of *Mycorrhiza*, *Azotobacter* and vermicompost increased plant height, number of leaves per plant, number of inflorescences per plant, stem diameter and fresh and dry matter, compared to control treatment.

Sandip (2015) reported that application of biofertilizer *Bacillus thuringiensis* and *Bacillus megaterium* resulted in high catalase and peroxidase activity in leaves, increased seed germination rate, sugar and phosphate content in leaves as well as dry weight.

Mostafa et al. (2011) observed that incolution with combination of three bacterial (*Pseudomonades sp.*, *Bacillus lentus*, *Azospirillum brasilens*) increased soluble carbohydrates (2.5µ mol glucose g^{-1} FW), chlorophyll (12 SPAD) and K content (250 mg. DW) in basil under water stress.

3.2. Effect of Bio-Fertilizers on Quality of *Ocimum basilicum*

Nazanin et al. (2014) conducted an experiment with four treatments viz., *Azotobacter chroococcum* (A) *Azospirillum lipoferum* (B) *Bacillus circulans* (C). The maximum geranial and the minimum caryophyllene in essential oil were obtained by using two biofertilizers (A + C). The highest methyl chavicol was obtained after applying two biofertilizers (B + C).

Shatta (2009) showed that the growth of basil plants and their active constituents were positively influenced by seedling inoculation with the asymbiotic N_2-fixers with organic fertilizer. Similarly, Hanan et al. (2010) indicated that application of 50% compost and 50% sand in the presence of biofertilizer resulted in enhancement of fresh and dry weights, total phenolics, total flavonoids and pigment content as compared with compost alone.

El-Naggar et al. (2015) reported that application of cattle manure at 142 m^3/ha combined with 2.0 and/or 4.0 g/l of biofertilizer resulted in significant increases in the main chemical components of leaves essential oil (estragole, eucalyptol, linalool, and trans-4-methoxycinnamaldehyde).

Sanjeet et al. (2016) indicated that the major constituent of essential oil (methyl chavicol) was significantly improved with application of (vermicompost with *Bacillus* sp.), (vermicompost with *P. monteilii*), and (vermicompost with *G. intraradices*) when compared to control.

Rashmi et al. (2008) studied the effects of inoculation of biofertilizers viz., *Glomus fasciculatum*, *Azotobacter chroococcum* and *Aspergillus awamori* singly or in combination on growth, biomass and biochemical constituents of *Ocimum gratissimum*. Inoculated plants showed increased plant height, number of leaves, number of branches, biomass, major and micronutrients, essential oil content and total phenol content.

Hanaa et al. (2016) reported that the highest percentages of Linalool, Camphor and Anethol were recorded in essential oil extracted from plants treated with combination of microorganisms encapsulated with sodium alginate, while the highest percentages of cineol resulted under the effect of combination of microorganisms carried on free suspension compared to control.

Baraa et al. (2017) revealed that bio-fertilizer has a good impact on oil quality, that highest percentage of Methyl chavicol was recorded with application of recommended FYM (10 t ha^{-1}) + recommended N through FYM along with bio fertilizers in the main crop (63.78%) and in the ratoon (59.39 and 59.67%) of 2015 and 2016, two years of experiments, respectively.

4. ECONOMICS OF CULTIVATION

The economic analysis of plant nutrient sources on sweet basil (*ocimum basilicum* L.) was evaluated by Thakur (2014). Net monetary returns were maximum with 100 percent recommended dose of fertilizer which fetched net income of Rs. 36,160/ha, the economics of production of tulsi has been worked out using farm-level data from the districts of Barabanki, Sitapur and Raebaerli in Uttar Pradesh. The net returns over total cost have been found to be 40,094 Rs/ha. The benefit-cost ratio however has been observed to be 3.21:1 (Ram et al., 2012).

The economic analysis for two years of experiment which has done by Baraa et al. (2017) and the results indicated that application of NPK (160:80:80 Kg/ha) + FYM (10 t/ha) recorded maximum net income (142,581.3 Rs. /ha), maximum gross income (184,371 Rs. /ha). While, the maximum B/C ratio (4.59) was with application of Rec. NPK (160:80:80 Kg/ha).

CONCLUSION

Intensive cropping systems with fertilizer responsive crops that rely on high input of inorganic fertilizers often lead to no sustainability in production and also pose a serious threat to soil health. Application of organic sources of nutrients along with recommended dose of inorganic fertilizers is rapidly being practiced. However, considering economics and also nutrient statue of the soil, entire dependence on organic sources of nutrients may not be adequate to attain the maximum productivity. In cultivation of basil, the real value is given to the quality while yield and quantity comes in the second step of importance. Sustainable agricultural approaches are the best methods in which this plant revealed better performance on the account of the harmony with nature. Therefore, current chapter critically discussed the optimum fertilization that lead to enhance the yield, give the best quality of basil oil and enrich the nutrient of the soil according to many research studies.

With knowing that Sweet basil (*Ocimum basilicum*) is a popular culinary herbal grown for fresh or dry leaf and essential oil. Recently, basil was shown to rank highest among species and herbal crops for phenolic compounds, which are associated with decreasing risks of cancer and aging diseases. So, this chapter has the potential to contribute information towards adoption best treatment that could increase the fresh and dry herb yield, oil yield and its best quality, ensure sustainability in agricultural systems as it relates to nutrient uptake in the soil and finally provide higher returns from money spent.

REFERENCES

Amiri, M.B., Dehghanipoor, F. and Tahami, M.K., 2013. Effects of Biofertilizers and Winter Cover Crops on essential oil production and some agroecological characteristics of basil (*Ocimum basilicum* L.) in an organic Agro-Systems. *J. Iranian Agro. Res.*, 10 (4): 751-763.

Anderson, S.H., Gantzer, C.J. and Brown, J.R. 1990. Soil physical properties after 100 years of continuous cultivation. *J. Soil and Water Cons.*, 45:117-121.

Anwar, M., Patra, D.D., Chand, S. and Kumar, A. 2005. Effect of organic manures and inorganic fertilizer on growth, herb and oil yield, nutrient accumulation, and oil quality of French Basil. *Communications in Soil Science and Plant Analysis*, 36(1): 1737–1746.

Arun, K.S., 2007. Biofertilizers for sustainable agriculture. Mechanism of P-solubilization. Sixth Edition. *Agri bios Publishers*, Jodhpur, India, 196-197.

Asieh, S., 2012 .Studying the effects of chemical fertilizer and manure on growth rate and essence amount of (*Ocimum basilicum* L.). *J. Eco. Environ.* 18 (3): 517-520.

Baraa, A., Suryanarayana M.A. and Kalaivanan, D. 2017. Effect of graded levels of N through FYM, inorganic fertilizers and biofertilizers on growth, herbage yield, oil yield and economics of sweet basil (*Ocimum basilicum* L.) *Medicinal Plants - International Journal of Phyto medicines and Related Industries*, 9(4):250.

Baraa, A., Suryanarayana M.A. and Kalaivanan, D. 2018. Influence of organic and inorganic fertilizers on yield and quality of sweet basil (*Ocimum basilicum* L.). *Journal of spices and aromatic crops,* 27(1):38-44.

Barea, J.M. and Brown, M.E. 2005. Effects on plant growth produced by *Azotobacter paspali* related to synthesis of plant growth regulating substance. *J. Appl. Bact.*, 37: 583-586.

Banerjee S., Palit, R., Sengupta, C. and Standing, D. 2010. Stress induced phosphate solubilization by Arthrobacter sp. and Bacillus sp. Isolated from tomato rhizosphere. *Aust. J. Crop Sci.*, 4(6): 378-383.

Beltrame, J., Angnes, R., Chiavelli, L., Costa, W., Montanher, S. and Rosa, M. 2014.Chemical composition of the essential oil obtained from *Ocimum basilicum* (Basil) cultivated in two regions from south Brazil. *J. Essential Oil-Bearing Plants*, 17(4):658-663.

Bot, A. and Benites, J., 2005. The Importance of Soil Organic Matter: Key to Drought-Resistant Soil and sustained Food Production (Bulletin No. 5). *Rome: Food and Agriculture Organization.*

Brix, H. 2008. *Soil Echangeable Bases (ammonium acétate method).* Available at : http://mit.biology.au.dk/biohbn/Protocol/Soil Exchangeable Bases CEC.pdf.

Daneshian, J., Yousefi, M., Zandi, P., Jonoubi, P. and Khatibani, L.B. 2011.Effect of planting density and cattle manure on some qualitative and quantitative traits in two basil varieties under Guilan condition, Iran. *American-Eurasian J. Agric. and Environ. Sci.,* 11(1): 95-103.

El-Naggar, A.H.M., Hassan, M.R.A., Shaban, E.H. and Mohamed, M.E.A. 2015. Effect of organic and biofertilizers on growth, oil yield and chemical composition of the essential oil of *Ocimum basilicum* L. Plants. *Alex. J. Agric. Res.*, 60 (1):1-16.

Ekwue, E.L. 1992. Effect of organic and fertilizer treatments on soil physical properties and erodibility. *Soil and Tillage Res.*, 22: 199-209.

Franchini, J.C. 2007. Microbiological parameters as indicators of soil quality under various soil management and crop rotation systems in southern Brazil. *Soil Res.*, 92(1): 18-29.

Gamal, E.B. 2004. Aromatic Plants of the Sudan. Ministry of Science and Technology. National center for Research. Medicinal and Aromatic Plants Research Institute, 2(3): 90- 111.

Gaur, A.C. and Gaind, S. 1990. *Role of phosphorus solubilizing microorganism in crop productivity and enriched organic manure. National seminar on organic farming org.* by JNKVV Jabalpur and IGKVV, Raipur at college of Agriculture Indore. 20-29 Sep. 1992 pp: 134-142.

Geetha, A., Rao, P.V., Reddy, D.V. and Shaik, M. 2009. Effect of organic and inorganic fertilizers on macro and micro nutrient uptake, oil content, quality, and herbage yield in sweet basil (*Ocimum basilicum*). *Res. Crops*, 10 (3):740-742.

Gundale, M.J. 2005.Restoration treatments in a Montana ponderosa pine forest: Effects on soil physical, chemical and biological properties. *Forest Eco. Manag.*, 213 (1): 25-38.

Hanaa, A., Abo-Kora and Maie, M.A. 2016. Reducing effect of soil salinity through using some strains of Nitrogen fixers bacteria and compost on sweet basil plant. *Int. J. Pharm Tech. Res.*, 9 (4):187-214.

Hanan, A., Taie, A., Zeinab, R. and Samir, R. 2010. Potential Activity of Basil Plants as a Source of Antioxidants and Anticancer Agents as Affected by Organic and Bio-organic Fertilization. *Not. Bot. Hort. Agrobot. Cluj*, 38 (1): 119-127.

Hussain, A. and Przybylski, R. 2008. Chemical composition, antioxidant, and antimicrobial activities of basil (*Ocimum basilicum*) essential oils depends on seasonal variations. *Food Chemistry*, 108(3): 986-995.

Jakowienko, P., Wójcik-Stopczyńska, B., AND Jadczak, D. 2011. Antifungal Activity of Essential Oils from Two Varieties of Sweet Basil (*Ocimum basilicum* L.). *Vege. Crops Res. Bulletin*, 74(1):25-33.

Jayasri, P. 2010. Effect of organic nutrients on growth and essential oil content of sweet basil (*Ocimum basilicum* L.). *J. Horti.*, 5 (1): 26-29.

Luiz, L., 2009. Organic fertilization in the production, yield and chemical composition of basil chemotype eugenol. *Hortic. Bras.*, 27(1): 37-40.

Kay, B.D. and Munkholm, L.J. 2004. Management induced soil structure degradation-Organic matter depletion and tillage. (in) *Managing Soil Quality: Challenges in Modern Agriculture*, pp 185–198. Schijonning P, Elmholt S and Christensen B T. (Eds), CABI, Wallingford, UK.

Kannaiyan, S. 2002. Biofertilizers for sustainable crop production. *Biotechnology of biofertilizers*, Narosa Publishing House, New Delhi, India. pp. 377.

Khosro, M., 2012. Bacterial biofertilizers for sustainable crop production: A review. *J. Agric. Bio. Sci.*, 7(5):307-3016.

Kuchenbuch, R.O. and Ingram, K.T. 2004. Effects of soil bulk density on seminal and lateral roots of young maize plants (*Zea mays* L). *J. Plant Nutria. Soil Sci.*, 167: 229–35.

Kumar, R., Bhatia, R., Kukreja, K., Behl, R.K., Dudeja, S.S. and Narula, N. 2007. Establishment of *Azotobacter* on plant roots: chemotactic response development and analysis of root exudates of cotton (*Gossypium hirsutum* L.) and wheat (*Triticum aestivum* L.). *J. Basic Microbiol.*, 47(5): 436-439.

Louie, G., Baiga, T., Bowman, M., Koeduka, T., Taylor, J. and Spassova, S. 2007. Structure and Reaction Mechanism of Basil Eugenol Synthase. *Plos ONE*, 2(10):993.

Martyniuk, S. and Martyniuk, M. 2003. Occurrence of *Azotobacter* spp. in some polish soils. *Polish J. Environ. Stud.*, 12(3): 371-374.

Marotti, R. 1996. Differences in Essential Oil Composition of Basil (*Ocimum basilicum* L.) Italian Cultivars Related to Morphological Characteristics. *J. Agric. Food Chem.*, 44 (1): 3926-3929.

Maryam, K., Mahdi, J. and Mostafa, M. 2013. Effect of vermicompost fertilizer on performance of Oil in Basil herb. *Intl. Res. J. Appl. Basic. Sci.*, 6 (7), 968-971.

Maryam, S. 2014. Effects of amino acids and nitrogen fixing bacteria on quantitative yield and essential oil content of basil (*Ocimum basilicum*). *Agric. sci. dev.*, 3(8) pp: 265-268.

Merestala, T., 1996. *Abstract Bibliography of researches of Bio and Organic Fertilizers at Benguet State University 1975-1996*. La Trinidad, Benguet. 53p.

Mohamad, F., Morteza, F., Elias, A. and Hossin, T. 2012. Effect of drought stress and types of fertilizers on the quantity and quality of medicinal plant Basil *(Ocimum basilicum* L.). *Indian J. Innovations Dev.*, 1 (9): 696-699.

Mostafa, H., Sayed, M. and Amir, G. 2011. Plant growth promoting rhizobacteria (PGPR) effect on physiological parameters and mineral uptake in basil (*Ociumum basilicm* L.) under water stress. *J. Agri. Bio. Scie.*, 6 (5):6-11.

Nazanin, R., Mohammad T. and Mohammadreza, H. 2014. Quantity and Quality of essential oil of basil (*Ocimum basilicum* L.) under biofertilizers application conditions. *Int. J. Adv. Biol. Biom. Res.*, 2(6):2134-2142.

Palada, M., Davis, A., Crossman, C., and Chichester, E.A. 2002. Sustainable crop management practices for improving production of culinary herbs in the Virgin Islands. In *XXVI International Horticultural Congress: The Future for Medicinal and Aromatic Plants*, 629: 289- 298.

Paramanik, R.C. and Chikkaswamy, B.K. 2014. Effects of VAM and Biofertilizers on some Medicinal Plants. *Int. J. Curr. Microbiol. App. Sci*, 3 (6): 1016-1027.

Randhawa, G.S., Gill, B.S. and Saini, S.S. 1998. Effect of Different Transplanting Dates and Crop Growth Stages on the Growth and

Physico-Chemical Properties of French Basil (*Ocimum basilicum*) Oil. *Indian Perfumer*, 38 (4): 123–128.

Raleigh, L. 2014. *Effect of organic nitrogen fertility on yield and nutritional content of basil and collard greens.* (PhD Thesis) Submitted to the Graduate Faculty of the University of Georgia.

Rashmi, K.R., Earanna, N. and Vasundhara, M. 2008. Influence of biofertilizers on growth, biomass, and biochemical constituents of *Ocimum gratissimum* L. *Biomed.*, 3(2): 123-130.

Salah, S. 2009. Herbal and essential oil yield of Genovese basil (*Ocimum basilicum* L.) grown with mineral and organic fertilizer sources in Egypt. *J. f-r. kulturp. flanzen.*, 61 (12): 443–449.

Samane, B. 2014. Changes in nitrogen and chlorophyll density and leaf area of sweet basil (*Ocimum basilicum* L.) affected by biofertilizer and nitrogen application. *J. Biosci.*, 5 (9): 256-265.

Sanjeet, K., Umesh, P., Khushboo, K., Khushboo, K., Rakshapal, S. and Rajesh, K. 2016. Bioinoculants and vermicompost improve *Ocimum basilicum* yield and soil health in a sustainable production system. *Clean Soil Air Water*, 44(6):686–693.

Sandip, B. 2015. Effect of dual inoculation of plant growth promoting rhizobacteria on different non-leguminous plants under pot condition, *Indian J. Microbiol. Res.*, 2(1):20-26.

Shaharoona, B., Arshad, M., Zahir, Z.A. and Khalid, A. 2006. Performance of Pseudomonas spp. containing ACC-deaminase for improving growth and yield of maize (Zea mays L.) in the presence of nitrogenous fertilizer. *Soil Biology and Biochemistry*, 38 (1):2971-2975.

Shatta, A.M., 2009.Effect of seedling inoculation with some asymbiotic N2-fixers on the growth of basil plant (*Ocimum basilicum*) and its active constituent. *J. Agri. Sci.*, 34 (3) pp: 1569 -1579.

Shirzadi, F., Ardakani, M. and Asadi Rahmani, H., 2014. Effect of vermicompost and biofertilizers on some quantitative characteristics of basil (*Ocimum basilicum* L.). *J. Agro.*, 6(3):678.

Shoae, 2013. Vermicompost, plant growth promoting bacteria can affect the growth and essence of basil (*Ocimum basilicum* L.). *Annals Bio. Res.*, 4 (2):8-12.

Simon, J.E., Morales, M.R., Phippen, W.B. and Vieira, R.F., 1999. A source of aroma compounds and a popular culinary and ornamental herb. In. J. Janick (Ed.): *Perspectives on new crops and new uses.* Alexandria, VA, ASHS Press. pp: 499-505.

Tahami Zarandi, S.M.K., Rezvan Moghaddam P. and Jahan M., 2010. Tasyrkvdhay comparison of organic and chemical yield and herbal essential oils of basil (*Ocimum basilicum* L.). *Journal of Agricultural Science Ecology*, 2: 74-63.

Tejera, N., Liuch, C., Martinez-Toledo, M.V. and Gonzalez-Lopez, J. 2005. Isolation and characterization of *Azotobacter* and *Azospirillum* strains from the sugarcane rhizosphere. *Plant Soil,* 270: 223-232.

Thakur, V., Sood, M., Mahajan, R. and Dutt, S. 2008. Effect of Organic Manure and Biofertilizers on Growth and Yield of Viola pilosa. *International Journal of Forest Usufructs Management*, 9 (1) 96-99.

Tuomisto, H.L., Hodge, I.D. and Riordan, P. 2012. Does organic farming reduce environmental impacts? A meta-analysis of European research. *J. Envi.Manag.*, 112: 309-320.

Theodore, J., 2012. *The response of basil (Ocimum basilicum L.) to chicken manure, compost, and urea applications.* (PhD Thesis) submitted to the graduate division of the University of Hawai'i in partial fulfillment of the requirements.

Vahid, M., Mohammad, R., Ardakani, F. and Avat, T., 2013. Physiological Responses of Sweet Basil (*Ocimum basilicum* L.) to Triple Inoculation with *Azotobacter*, *Azospirillum*, *Glomus intraradices* and Foliar Application of Citric Acid. *Annals Bio. Res.,* 4 (1):62-71.

Ved, D.K. and Goraya, G.S., 2008. Demand and Supply of Medicinal Plants in India. Bishan Singh Maheandra Pal Singh, pp: 70.

Violen, H.G.M., 2007. Alternation of tomato fruit quality by root inoculation with plant growth promoting rhizobacteria (PGPR). *Scientia Hortic.*, 113 (1): 103-106.

Wu, C, Cao, Z.H., Li, Z.G. and Cheung, K.C., 2005. Prevalence of beta proteobacterial sequences in gene pools associated with roots of modern rice cultivars. *Geoderma*, 125(1): 155-166.

Zarghari, A., 1997. *Medicinal Plants*. Tehran University Publications, Iran, p: 921.

BIOGRAPHICAL SKETCH

Baraa ALmansour, PhD

Affiliation: PhD holder from University of Horticultural Sciences, Bagalkot -India

Education: PhD in Horticulture (Medicinal and Aromatic plants)

Research and Professional Experience: Organic cultivation, Aromatherapy

Publications from the Last 3 Years:

1. Effect of graded levels of N through FYM, inorganic fertilizers and biofertilizers on growth, herbage yield, oil yield and economics of sweet basil (*Ocimum basilicum* L.) *Medicinal Plants - International Journal of Phyto medicines and Related Industries*, 2017: 9(4):250. ·
2. Effect of integrated nutrient management on dry herbage yield, nutrient uptake and profitability of French Basil (*Ocimum basilicum* L.). *Journal of Horticultural Science*, 2017:12(2):171-179.
3. Influence of organic and inorganic fertilizers on yield and quality of sweet basil (*Ocimum basilicum* L.). *Journal of spices and aromatic crops*, 2018:27(1):38-44.

4. Effects of organic and inorganic fertilizers on soil fertility, nutrient uptake and yield of French basil. *International Journal of Phyto medicines and Related Industries*, 2019:11(1):40-45.
5. Effect of nutrient integrated management on oil yield, nitrogen balance and economics of Basil (*Ocimum basilicum*) cultivation. *Asian Journal of Biological Sciences*, 2019: 12: 750-757.
6. Nutrient Uptake, Post - Harvest Nutrient Availability and Nutrient balance Sheet under Integrated Nutrient Management Practices in Sweet Basil (*Ocimum Basilicum* L.) Cultivation. *Global Journal of Science Frontier Research: D Agriculture and Veterinary*, 2020: 20(3): 1-6.
7. Traditional Medicine "Ayurveda" Guidelines against COVID-19 Addressing Ayush Ministry Protocol. *Journal of Natural & Ayurvedic Medicine,* 2020:4(2):1-4.
8. Concept of Aromatherapy in boosting psychological immune system against COVID-19. *Medicinal Plants, 2020* 12(2):161-168.
9. Essential Oil Content and Composition of Sweet Basil (*Ocimum basilicum.* L) Under Integrated Nutrient Management. *Global Journal of Agricultural Innovation, Research & Development*, 2019: 6- 25-31.

In: *Ocimum basilicum*
Editor: Andres A. Walton

ISBN: 978-1-53619-265-0
© 2021 Nova Science Publishers, Inc.

Chapter 3

AN OVERVIEW OF *O. BASILICUM* (L.) IN TURKEY

*Muhammad Azhar Nadeem[1], Yeter Çlleslz[1],
Ecenur Korkmaz[2], Zemran Mustafa[3],
Faheem Shehzad Baloch[1], Tolga Karaköy[1]
and Muhammad Aasim[1],**

[1]Department of Plant Protection,
Faculty of Agricultural Sciences and Technologies,
Sivas University of Science and Technology, Sivas, Turkey
[2]Department of Agricultural Sciences,
Faculty of Agricultural Sciences and Technologies,
Sivas University of Science and Technology, Sivas, Turkey
[3]Department of Plant Production Technology,
Faculty of Agricultural Sciences and Technologies,
Sivas University of Science and Technology, Sivas, Turkey

---

* Corresponding Author's E-mail: maasim@gmail.com.

ABSTRACT

Ocimum basillicum L. sweet basil is an annual herbaceous plant belonging to *Ocimum* genus of the Lamiaceae family. The genus *Ocimum* contains 50 to 150 herb and shrub species and shows the natural distribution in Asia, Africa, and Central America. Sweet basil has a good concentration of vitamins and minerals and potentially used as a medicinal plant for the treatment of various diseases. Sweet basil reflects great variations in terms of its morphology and chemical contents and is mainly grown as a spice and ornamental plant. The essential oil contents of sweet basil vary 0.5% to 1.5% depending on the climatic conditions. Additionally, sweet basil is considered a good source of various phenolic compounds that have beneficial effects on human health. Sweet basil is also used as spice, medicine, food, and perfumery industries. The essential oil of this plant has increasing importance due to various biological effects such as antifungal, insecticide, and antioxidant. In addition, the purple-colored varieties of sweet basil are an important anthocyanin source for the food industry. In Turkey, sweet basil is widely used for culinary and medicinal purposes and its leaves are used as salad. There is a number of registered cultivars of sweet basil in Turkey that are used for different purposes ranging from food to medicine and ornamental plant. This study summarized the potential uses, cultivation, and available germplasm of *O. basilicum* in Turkey.

Keywords: sweet basil, ornamental, medicinal, culinary herb, spice

1. INTRODUCTION

The genus *Ocimum* belongs to the Lamiaceae family and its individuals have wide distribution covering Asia, Africa, Europe, and Central American continents (Alves-Silvaa et al. 2013; Ekmekci and Aasim 2014). This genus contains 50 to 150 species (Simon et al. 1999). The most common and well-known medicinal and culinary herb plant of ocimum genus is sweet basil (*O. basilicum*) (Siddiqui et al. 2012). Sweet basil is generally native to the Indian Sub-continent but also found in different parts of the World (Purushothaman et al. 2018; Aasim 2020). Due to its wide distribution, it exhibits morphological (Nurzyńska-Wierdak

2007) and chemical/biological variations (Nguyen et al. 2010; Aasim 2020). In most parts of the World, it is cultivated as a culinary herb (Gulcin et al. 2007), medicinal (Bora et al. 2011), ornamental (Javanmardi et al. 2003), edible plant (Makri and Kintzois 2008), and for obtaining by-products for variable purposes (Nguyen et al. 2010).

The sweet basil plant is a multi-purpose cultivated perennial herb (Bantis et al. 2016). It is used as a whole or its parts are used in different forms for various traditional medicinal systems (Purushothaman et al. 2018) like treating face pimples in Indian Siddha medicine (Tsai et al. 2011). Moreover, the usage of sweet basil varies according to objectives and region. For example, the infusions are practiced for shrinking the plasma lipid content in Mediterranean areas (Bravo et al. 2008), diabetics (Rai et al. 2009) and cardiovascular disorders (Umar et al. 2014) in Turkey, aches, and pains in Bulgaria (Opalchenova et al. 2003) and sedative in Spain (Freire et al. 2006). According to Purushothaman et al. (2018), more than 20 different compounds are present in sweet basil leaves that exhibits stimulating and antispasmodic effects. The usage of sweet basil has been found beneficial in relieving headaches, stomach cramps, and digestive problems. Moreover, it has beneficial effects in relieving colic pain and swelling. It has an increasing effect on breast milk as well. The basil juice or scent repels fly bites, mosquitoes and insects like ticks (Del Fabbro and Nazzi 2008). The seeds have an effect that soothes the nerves, but also helps with depression and tension. Basil seeds or essential oil in combination with other essential oil sources have been reported synergistic effects with relatively high efficiency for curing insomnia symptoms (Jezdic et al. 2018). Some other uses of basil include anticonvulsant (Freire et al. 2006; Nguyen et al. 2010), antifungal (Dambolena et al. 2010), anti-hyperlipidemic (Amrani et al. 2009), anti-inflammatory (Raina et al. 2016), anti-microbial (Nguyen and Niemeyer 2008), antioxidant (Pandey et al. 2016), antiplatelet property (Amrani et al. 2009), anti-thrombotic (Tohti et al. 2006), cardiotonic and abdominal pain reliever (Bais et al. 2002), coccidial activity in the broiler chicks (Kumar et al. 2011), cytotoxicity effect (Aarthi et al. 2010), immunomodulatory (Okazaki et al. 1998) and insecticidal activities (Freire et al. 2006).

Previous studies have confirmed phytochemical properties in sweet basil. Sweet basil is used for other purposes like; (i) basil soap, (ii) basil tea (iii) basil oil, (iv) smoothies, (v) milkshakes, (vi) lemonade and other drinks, (vii) soups, (viii) salad dressings, (ix) yoghurt (x) pudding (xi) rolled oats, (xii) whole grain pancakes and (xiii) pastas and (xiv) bread and other pastry products (Anonymous 2020a). Although sweet basil usage in a different form is considered to be safe, the presence of alkenyl benzenes in some chemotypes may lead to toxic impact (Sestili et al. 2018). The possible negative or toxic impact of basil may include inconvenience for pregnant and breastfeeding women and should not be used without consulting a doctor. It may cause a blood sugar lowering effect when used as a raw form. It may be inconvenient for people with low blood pressure to use it. It is recommended that people who will undergo surgery should not consume basil until at least a week before, as it has an increasing effect on bleeding.

Basil likes sandy-loamy soils rich in organic matter, deep structured, and high-water holding capacity. Although it is sensitive to cold, it is resistant to heat as well. It likes the sun and rainy climates. It's sowing starts in autumn in coastal regions with a mild climate and without winter frosts, and in spring in regions with harsh winters. Seeds are widely planted and can be sown directly in the field using a seed drill. To achieve this, almost 5-10 kg seeds considered sufficient per hectare (ha). Depending on the variety used, the row spacing should be between 20 and 60 cm. After emergence, thinning should be performed to keep 10 - 30 cm between plants. The plant forms a large number of branches. These branches stop growing when flowering begins, and they become wood quickly. To prevent this, shoots with buds are broken at the bottom and the plant is encouraged to produce leaves. Its cultivation is done in pots and garden in Turkey and the preferred time is May and it requires more water in the initial life cycle (Anonymous 2020a).

2. *O. BASILICUM* IN TURKEY

O. basilicum is a relative of mint, thyme, and wild thyme and is known as "Fesleğen" or "Reyhan" in the Turkish language. The natural occurrence of basil in Turkey is not reported especially in Western and Southern Anatolia. In these regions, it is mostly grown in pots in the house and garden and even in a balcony. In Eastern Provinces, purple-colored types are common than other types and are known as basil. Green colored varieties are more common in Western provinces and known as "sweet bacillus." Although its homeland is South Asia, it is often grown in the Aegean and Mediterranean coastal regions of Turkey. Sweet basil is mainly cultivated in Izmir, Hatay, Ordu and Isparta provinces of Turkey (Anonymous 2020b). There are number of local and exotic basil cultivars and genotypes available for commercial production in some parts of Turkey. The purpose of cultivated basil in Turkey is for seasoning (Günay and Telci 2017), spice, and ornamental plants (Nacar 1997). However, the cultivated area is very low with the low land cultivated unit and hence, there is no data available about its cultivated area and production at local and international level.

The most common basil types cultivated in the World and Turkey with general characteristics are given below (Anonymous 2020c).

a) *Cinnamon Basil:* It is known as micro basil. Harvesting is performed before flowering.
b) *Spicy Bush Basil*: It is used before the leaves are separated from the stem and fully mature.
c) *Opal Basil*: It is dark purple in color having a flavor of cloves. It is generally used for desserts, sauce-syrups. and cocktails.
d) *Mixed Bud Basil*: It is aromatic and tastes like licorice. It has purple color and often used in sauces.
e) *Thai Basil*: Its aroma is aniseed. This taste is not lost even when the plant is heated. It is used as a scented sauce type.

f) *African Blue Basil*: It tastes aniseed and spicy. It is used as a sauce in meat and chicken dishes.
g) *Ceneviz Basil*: It tastes sweet and is used as a sauce.
h) *Lemon Basil:* It tastes sour and is used in sauces.

3. GENOTYPES/LANDRACES/POPULATIONS OF *O. BASILICUM* IN TURKEY

Morphological studies on basil revealed more than 80 local basil genotypes from different regions in Turkey based on agronomic and other characterization. It was investigated that there is a wide morphological and chemical variation in local basil germplasm, and some genotypes have been stated to stand out in terms of yield and essential oil content. Sweet basil reflects a great level of diversity for various chemical components (linalool, citral, methyl cinnamate, methyl eugenol, etc.). To date, number of variable genotypes (G) and landraces (LR) of sweet basil has been reported. However, the relative information about these G/LR is limited even not available in the local language (Turkish). The relative information available about G/LR is the coding based on geographical distribution with no information about their characteristics. A good number of research studies have been done and published in both Turkish and English languages by considering herb yield and essential oil contents of this plant. To exploit the potential of these local G/LR, conservation and commercial production are essential to estimate the yield and essential oil production based on the different geographical distribution (Telci et al. 2005). Some of the G/LR with their general characteristics are given below:

- *Adana Normal Leafy Basil*: Stems and leaves are green, flowers are white, calyx glabrous and calyx teeth are equal in length (Erşahin and Özgüven 2010).
- *Adana Small Leaved Basil*: The branches and leaves are green, with white flowers, the calyx hairs are not equal in length, and the

two females are longer than the others (Erşahin and Özgüven 2010).
- *Antalya*: It is short, fleshy, and has large leaves, and its trunk and branches are well developed. Leaves are dense at the bottom (Telci 2005).
- *Diyarbakir Purple Basil*: Branches and leaves are purple, flowers are pink, calyx is sparsely hairy, and calyx teeth are of equal length (Erşahin and Özgüven 2010).
- *Diyarbakir Top Basil*: Branches and leaves are green, flowers are white, calyx is sparsely hairy, and calyx teeth are not of equal length (Erşahin and Özgüven 2010).
- *Gaziantep*: The leaves are small, smooth and green in color, while flowers are white. The length of branches are medium in size and semi-horizontal (Cabar 2016).
- *İzmir-1*: Branches and leaves are green, flowers are white, calyx is sparsely hairy and calyx teeth are of equal length (Erşahin and Özgüven 2010).
- *İzmir-2*: Branches and leaves are mottled purple, flowers are pink, calyx is sparsely hairy and calyx teeth are of equal length (Erşahin and Özgüven 2010).
- *Kırşehir*: The leaves are medium-sized, smooth-surfaced and green in color. The flowers are white in color. It has few branches and develops semi-horizontally (Cabar 2016).
- *Mersin*: It is a high stature, large habitus genotype. The fragrant is like Melissa. It is tall with large habitus and its branching rate is lower than others. There are prominent and dense hairs on calix and brackled leaves. It is more important for the essential oil industry than for use as a spice (Telci 2005)
- *Osmaniye*: Branches and leaves are mottled purple, flowers are light pink, calyx glabrous or sparsely hairy, calyx teeth of equal length (Erşahin and Özgüven 2010).

- *Şanlıurfa*: The leaves are small, smooth and green in color. Flower color is white. It has many branches and develops semi-horizontally (Cabar 2016).
- *Zonguldak*: Morphologically, it is very similar to the varieties known as "sweet bacillus" in literature. But it is rich in methyl chavicol and calyx is dark colored. Efficient and high in essential oil (Telci 2005).

4. CHARACTERIZATION OF ESSENTIAL OIL AND YIELD OF LOCAL LANDRACES OF TURKEY

A good number of studies has been conducted aiming to explore herb yield and essential oil contents using local sweet basil germplasm (genotypes/landraces) collected from various geographical regions of Turkey. These regions are the Aegean, Black Sea, Mediterranean, and South-Eastern Anatolian region. Telci et al. (2006) used Turkish basil germplasm and revealed seven different chemotypes named (i) Linalool, (ii) Methyl cinnamate, (iii) Methyl cinnamate/linalool, (iv) Methyl eugenol, (v) Citral, (vi) Methyl chavicol (estragole) and (vii) Methyl chavicol/ citral. They also reported the "new chemotype basil" which was rich in Methyl chavicol with high citral contents. The characterization of basil essential oil of 14 different Turkish basil genotypes (Zonguldak, Antalya, Tokat, Malatya, Adıyaman, Denizli, Anamur, Burdur, Gaziantep) obtained from different regions of Turkey were compared with commercial varieties of France origin. The Turkish genotypes expressed the remarkable chemical variability from the essential oil composition. The chemical profiling of 10L and 17 genotypes constituted 42.17% and 44.80% more citral and 30.56% and 32.03% methyl chavicol. It was also notable that Citral/methyl chavicol was assessed as a new chemotype from cultivated basil in Turkey. Previous study grouped basil genotypes into two major clusters based on RAPD analysis along with chemical characterization with very few exceptions (genotype no. 6). A correlation analysis expressed low values (r = -0.40) for the genetic distance matrix and Euclidian distance matrix (Giachino et al. 2014).

4.1. Essential Oil and Yield Characterization of Local Landraces under Mediterranean Region

The effect of row distances on basil genotypes originated from different provinces of Turkey (Osmaniye, Adana, Hatay, Kahramanmaraş, Greece and France) was investigated under Adana climatic conditions. The results revealed 25000-30000 kg/ha green herb, 5000-7500 kg/ha dry herb and 1200-2000 kg/ha dry leaf. The impact of air temperature was highly significant and found to be effective on genotype efficiency (Nacar 1997). In another study under the Mediterranean region of Turkey using different basil varieties from France, Greece, Germany and Turkey, essential oil components were highly associated with cultivars and harvesting time. The results also revealed three different chemotypes based on compounds like linalool (Greek basil), linalool+methyl eugenol (French basil) and linalool+methyl cinnamate (Turkish and German basil) (Nacar and Tansı 2000). Tansı and Nacar (2000) used commercially cultivated lemon basil (*O. basilicum* var Citriodorum) under Çukurova conditions and harvested three times at the full flowering stage. The results revealed total fresh leaf yield (27600 kg/ha) and total dried leaf yield (5715 kg/ha) with maximum oil contents of 0.71% at first harvest. The main content was citral (neral+geraniol) containing 89.28% at the second harvest. The typical Mediterranean climate includes cool and rainy winters and hot and dry summers. Günay (2017) found the average plant height between 30.6-65.2 cm in 2015 and 36.1-75.3 cm in a study conducted with 10 different basil genotypes for two years in Isparta ecological conditions. The results highlighted the highest green herb yield (50416 kg/ha), dry herb yield (7981 kg/ha), dry leaf (4538 kg/ha) and essential oil ratio varied between 0.11-1.45%.

Günay and Telci (2017) conducted a study during 2015 and 2016 to determine the yield and quality characteristics of basil genotypes selected from Isparta ecology. The yield and essential oil changes in the region of 10 linalool and citral rich genotypes selected from previous studies were examined. The total yield was determined in genotypes rich in citral during the year, and found that the green herb yield increased up to 5.0 tons/da during the study. Essential oil yield increased up to 2.6 L/da. The highest linalool ratio was determined in R19 genotype with 76.7-79.3%. Change intervals in citral ratios were high and the highest value in citral rich types was obtained from R17 with 73.6%. As a result of this study; It has been determined that citral-rich types can be produced as essential oil and linalool-rich types as spice for citral production, which is especially preferred in cosmetics in the region.

Sönmez et al. (2019) aimed to explore the effects of different growth times on yield and quality characteristics of green and purple basil (*O. basilicum* L.) types. During their study, various plant growth parameters like plant height (cm), green herb yield (kg da^{-1}), herb yield (kg da^{-1}), drug leaf yield (kg da-1), essential oil ratio (%) and essential oil yield (1 da-1) properties were investigated. They resulted that maximum mean values for evaluated parameters were obtained in green basil and mostly in the 2nd, 3rd and 4th form times.

4.2. Essential Oil and Yield Characterization of Local Landraces under Aegean Region

Alonso et al. (1995) collected sweet basil germplasm from the Fethiye region of Turkey aiming to investigate the essential oil and reported 30 different components in the studied germplasm. The main components identified in their study were linalool (43.75%), trans-methyl cinnamate (4.63%) and classified as linalool-methyl cinnamate chemotype. Arabaci and Bayram (2004) used to investigate the effect of various nitrogen doses (0 and 50 kg/ha) and plant densities (20x20, 40x20 and 60x20 cm) on basil yield for three years in Aydın (Aegean) ecological conditions. Their results revealed 30070-42657 kg/ha, 9110-10070 kg/ha drog herb yield, 4701-6680 kg/ha drog leaf yield and 0.62-1.00% essential oil ratio (Arabaci and

Bayram 2004). A study by Erken et al. (2009) aimed to explore the effect of different planting densities on the yield and quality characteristics of the basil plant grown in Bornova ecological conditions for two consecutive years. Their results revealed a decrease in plant height, green herb yield and drog herb yield every 2 years. In both years, the highest results in terms of green herb yield, drug-herb yield and drug leaf yield were achieved at a planting density of 20x10 cm. In terms of essential oil ratio, they observed highest values from 40x10 cm in the first year and 30x10 cm in the second year.

In a study conducted by Karık et al. (2014) to determine the morphological, yield and quality characteristics of commercial and local basil cultivars in Menemen ecological conditions. They found that local varieties, "Kırmızı" and "Anamur" were ahead of commercial varieties in terms of yield characteristics. The local variety "Anamur" was the leader in terms of essential oil ratio and essential oil yield among commercial and local varieties. Moreover, they also reported 23 components in essential oils. The main components and proportions of essential oils were linalool (0.41-74.43%), 1.8-cineole (8.47-44.94%) and p-allylanisole (5.22-22.53%), respectively (Karık et al. 2014). Aslan (2014) used a total of six basil genotypes and the experiment was conducted in Aydın ecological conditions. During his study, he found average plant height (37.64-95 cm), green herb yield (7953-35767 kg/ha) and drug-herb yield (2371-12250 kg/ha), drug leaf yield (97.92-542.42 kg/ha) and average drug flower yield (52.08-339.830 kg/ha). The essential oil ratio in the flower and leaf varied between 0.13-1.23% and 0.18-1.70% respectively. The main essential oil components were methyl chavicol and eugenol.

4.3. Essential Oil and Yield Characterization of Local Landraces under Black Sea Region

Telci et al. (2005) conducted a study at Ordu University Faculty of Agriculture, Field Crops Department during 2014 to determine the herb yield and essential oil ratio of some basil populations. They used

Coincidence Blocks Trial Pattern with three replications and plant material was collected from different parts of Turkey. The experimental material was comprised of 9 basil genotypes, originating from Gaziantep (G1, G3, G8, G9), Tokat (G2), Sivas (G4), Adana (G5), Yozgat (G6) and Antalya (G7) provinces. It was determined that all examined traits showed statistically significant differences among the genotypes. The difference between harvest times was not significant for any studied trait. Genotype x form time interaction was found significant only at plant height. As a result of the study, plant height (17.16-45.33 cm), green herb yield (195.00-383.99 g/plant), dry herb yield (22.21-46.85 g/plant), dry leaf herb yield (12.46-25.99) was recorded in 9 basil genotypes. The essential oil ratio (g/plant) varied 0.25-1.06%.

Phenological, morphological and quality characteristics of four basil populations were examined in 2005 under Samsun (Black sea) conditions. The attained results were fresh herb yield of 7590-18660 kg/ha, drog herb yield of 903-8414 kg/ha, drog leaf yield of 351- 4483 kg/ha, dry leaf ratio of 14.3-68.28%, plant height of 7-47 cm and essential oil ratio of 0.35-0.95%. In terms of average values, fresh herb yield (14000 kg/ha), drog herb yield (2822 kg/ha), drog leaf yield (1637 kg/ha), dry leaf ratio (53.75%), plant height (22.87 cm) and essential oil ratio 0.74% were recorded (Uzun 2009). The study conducted under Ordu ecology revealed plant height (22.65-64.13 cm), dry leaf ratio (37.56-68.50%), green herb yield (5622-19982 kg/ha), dry herb yield 729-2352 kg/ha) and 0.14-1.53% essential oil rate (Özcan 2014). Another study conducted in Ordu with 9 different basil genotypes and results per plant revealed plant height 17.16-45.33 cm, green herb yield (195.00-383.99 g), dry herb yield (22.21-46.85 g), dry leaf herb yield (12.46-25.99) and essential oil varied between 0.25-1.06% (Karaca 2017). Comparison of genotypes revealed highest total green herb yield were recorded from Tokat origin G2 genotype. Gaziantep origin G1 genotype yielded maximum total dry herb yield and Tokat origin G2 genotype yielded total dry leaf herb yield. In general, the G2 genotype of Tokat origin has the highest yield in terms of green herb yield and dry leaf herb yield. On the other hand, among the genotypes, the Adana origin

G5 genotype draws attention with the highest essential oil content (Karaca et al. 2017).

4.4. Essential Oil and Yield Characterization of Local Landraces under Eastern and South Eastern Anatolian Region

Özek et al. (1994) conducted a study in Gaziantep province aiming to explore the the essential oil composition in sweet basil. They investigated a total of 60 components with major components of linalool (24%), E-methyl cinnamate (16.72%) and 1.8-cineol (13.63%). Özcan and Calchat (2002) investigated the essential oil composition in two locally grown basil species (*O. basilicum* L. and *O. minimum* L.) collected from Ovacik-Gülnar. Their results revealed 88.1% and 74.4% essential oil in *O. basilicum* and *O. minimum* respectively. The main components found in sweet basil L. were 78.02% methyl eugenol, 6.17% α cubebin, 0.83% nerol 0.74% ε-murolene. Whereas, the main components of *O. minimum* were 69.48% geranyl acetate, 2.35% terpinen-4-ol and 0.72% octan-3-yl-acetate. Erşahin (2006) set up an experiment at 30x70 cm planting density to determine the yield, and quality characteristics of the Adana, Osmaniye, Izmir, and Diyarbakır basil populations in Diyarbakır ecological conditions. The results obtained were plant height (37.13-82.07 cm), branch number (10.67-27.47), green herb yield (4210-31970 kg/ha), drog herb yield (784-6441 kg/ha), drog leaf yield (547- 3393 kg/ha) and the essential oil ratio of 0.49-1.25% in drog leaves. Adana and Diyarbakır populations were superior in yield and quality compared to other populations.

CONCLUSION

Traditionally, sweet basil has been used as a condiment and as a folk remedy for the treatment of diseases. It is a good source of terpenoids, alkaloids, flavonoids, tannins, saponin glycosides and ascorbic acid. In Turkey, sweet basil is used as a folk medicine and traditional Uyghur medicine to prevent and treat diabetics and cardiovascular disorders. Aegean, Black Sea, Mediterranean, and South-Eastern Anatolian regions are considered main growing areas of sweet basil. These regions contain a good number of landraces that have been used for the identification of essential oil and yield related traits. As there are a good number of landraces in Turkey, efforts should be made to develop sweet basil cultivars regarding various traits of interest.

REFERENCES

[1] Aarthi, N. and Murugan, K. (2010). Larvicidal and repellent activity of *Vetiveria zizanioides* L, *Ocimum basilicum* Linn and the microbial pesticide spinosad against malarial vector, *Anopheles stephensi* Liston (Insecta: Diptera: Culicidae). *Journal of Biopesticides*. 3: 199–204.

[2] Aasim, M. (2020). An overview on in vitro regeneration and genetic transformation in sweet basil (*Ocimum basilicum* L.). In: *Ocimum an Overview* (Ed) Blanchard M. Nova Science Publishers Inc. USA. pp47-84.

[3] Perez-Alonso, M. J., Velasco-Negueruela, A., Duru, M. E., Harmandar, M., and Esteban, J. L. (1995). Composition of the essential oils of *Ocimum basilicum* var. *glabratum* and *Rosmarinus officinalis* from Turkey. *Journal of Essential Oil Research*, 7(1), 73-75. https://doi.org/10.1080/10412905.1995.9698467

[4] Alves-Silvaa, J. M., Dias dos Santos, S. M., Pintado, M. E., Perez-Alvarez, J. A., Fernandez-Lopez, J. and Viuda-Martos, M. (2013). Chemical composition and in vitro antimicrobial, antifungal and antioxidant properties of essential oils obtained from some herbs widely used in Portugal. *Food Control*. 32(2): 371–378. https://doi.org/10.1016/j.foodcont.2012.12.022.

[5] Amrani, S., Harnafi, H., Gadi, D., Mekhfi, H., Legssyer, A., Aziz, M., Martin-Nizard, F. and Bosca, L. (2009). Vasorelaxant antiplatelet aggregation effects of aqueous *Ocimum basilicum* extract. *Journal of Ethnopharmacology.* 125(1): 157–62. https://doi.org/10.1016/j.jep.2009.05.043 PMid:19505553.

[6] Anonymous (2020a). *Medicinal and aromatic plants* https://ankara.bel.tr/files/8815/6499/5402/FESLEEN.pdf.

[7] Anonymous (2020b). http://www.tubives.com/.

[8] Anonymous (2020c). *Basil types and cultivation.* http://eceyda.com/feslegen-cesitleri-ve-yetistiriciligi/.

[9] Arabaci, O., and Bayram, E. (2004). The effect of nitrogen fertilization and different plant densities on some agronomic and technologic characteristic of *Ocimum basilicum* L. (Basil). *Journal of Agronomy.* 3(4): 255-262.

[10] Aslan, D. F. (2014). *Farklı Reyhan (Ocimum basilicum L.) genotiplerinde ontogenetik ve morfogenetik varyabilitenin belirlenmesi* (Master's thesis, Adnan Menderes Üniversitesi, Fen Bilimleri Enstitüsü). [*Determination of ontogenetic and morphogenetic variability in different Reyhan (Ocimum basilicum L.) genotypes*]

[11] Bais, H. P., Walker, T. S., Schweizer, H. P. and Vivanco, J. M. (2002). Root specific elicitation and antimicrobial activity of rosmarinic acid in hairy root cultures of *Ocimum basilicum*. *Plant Physiology and Biochemistry.* 40(11): 983–995. https://doi.org/10.1016/S0981-9428(02)01460-2.

[12] Bantis, F., Ouzounis, T. and Radoglou, K. (2016). Artificial LED lighting enhances growth characteristics and total phenolic content of *Ocimum basilicum*, but variably affects transplant success. *Scientia Horticulturae.* 198: 277–283. https://doi.org/10.1016/j.scienta.2015.11.014.

[13] Bravo, E., Amrani, S., Ariz, M., Harnafi, H. and Napolitano, M. (2008). *Ocimum basilicum* ethanolic extract decreases cholesterol synthesis and lipid accumulation in human macrophages. *Fitoterapia.*

79(7-8): 515–523. https://doi.org/10.1016/j.fitote.2008.05.002 PMid: 18620033.

[14] Cabar, B. S. (2016). *Determination of some yield and quality components of sweet basil (Ocimum Basilicum L.) lines from different origins in Thrace region.* Master's Thesis. Yüksek Lisans Tezi.

[15] Bora, K. S., Arora, S. and Shri, R. (2011). Role of Ocimum basilicum L. in prevention of ischemia and reperfusion-induced cerebral damage, and motor dysfunction in mice brain. *Journal of Ethnopharmacology.* 137(3): 1360-1365. doi: 10.1016/j.jep.2011. 07.066 Pmid: 21843615.

[16] Dambolena, J. S., Zunino M. P., López, A. G., Rubinstein, H. R., Zygadlo, J. A., Mwangi, J. W., Thoithi, G. N., Kibwage, I. O., Mwalukumbi, J. M. and Kariuki, S. T. (2010). Essential oils composition of *Ocimum basilicum* L. and *Ocimum gratissimum* L. from Kenya and their inhibitory effects on growth and fumonisin production by *Fusarium verticillioides. Inn. Food. Sci. Emerg.* Technol. 11(2):410–414.

[17] Del Fabbro, S. and Nazzi, F. (2008). Repellent effect of sweet basil compounds on *Ixodes ricinus* ticks. *Experimental and Applied Acarology.* 45(3-4), 219-228.

[18] Ekmekci, H. and Aasim, M. (2014). *In vitro* plant regeneration of Turkish sweet basil (*Ocimum basilicum* L.). *Journal of Animal and Plant Sciences.* 24(6): 1758-1765. ISSN: 1018-7081.

[19] Erken, S., Sönmez, Ç., Sancaktaroğlu, S. and Bayram, E. (2009). The effect of different planting densities on the yield and quality characteristics of basil (*Ocimum basilicum* L.) plant. *Ege University Faculty of Agriculture Journal, 46*(3), 165-173.

[20] Erşahin, L. and Özgüven, M. (2010). *Agronomic and quality characteristics of basil (Ocimum Basilicum L.) populations grown in Diyarbakir ecological conditions.* Çukurova University Institute of Science and Technology Year: 2010 Skin:22-2.

[21] Erşahin, L. 2006. *Agronomic and quality characteristics of basil (Ocimum basilicum L.) populations grown in Diyarbakır ecological*

conditions. Master Thesis, Çukurova University, Graduate School of Science, Field Crops Department, Adana.

[22] Freire, C. M. M., Marques, O. M. M. and Costa, M. (2006). Effects of seasonal variation on the central nervous system activity of Ocimum gratissimum L. essential oil. *Journal of Ethnopharmacology.* 105(1-2): 161–166. Doi: 10.1016/j.jep.2005.10.013 Pmid: 16303272.

[23] Giachino, R. R. A., Sönmez, Ç., Tonk, F. A., Bayram, E., Yüce, S., Telci, I. and Furan, M. A. (2014). RAPD and essential oil characterization of Turkish basil (*Ocimum basilicum* L.). *Plant systematics and evolution.* 300(8), 1779-1791.

[24] Gulcin, I., Elmastas, M. and Aboul-Enein, H. Y. (2007). Determination of antioxidant and scavenging activity of Basil (Ocimum basilicum L. Family Lamiaceae) assayed by different methodologies. *Phytotherapy Research.* 21(4): 354–361. https://doi.org/10.1002/ptr.2069 PMid:17221941.

[25] Günay, E. and Telci, I. (2017). Determination of yield and quality characteristics of some reyhan (*Ocimum basilicum* L.) genotypes in Isparta ecological conditions. *Journal of the Faculty of Agriculture.* 12(2), 100-109.

[26] Günay, E. (2017). *Determination of yield and quality characteristics of some reyhan (Ocimum basilicum L.) genotypes in Isparta conditions.* Master Thesis. Süleyman Demirel University. Institute of Science. Field Crops Division. 61s. Isparta.

[27] Javanmardi, J., Stushnoff, C., Locke, E. and Vivanco, J. M. (2003). Antioxidant activity and total phenolic content of Iranian *Ocimum* accessions. *Food Chemistry.* 83(4): 547–50. https://doi.org/10.1016/S0308-8146(03)00151-1.

[28] Jezdic, Z., Ozimec-Vulinac, S., Racz, A., Kovacevic, I., Sedic, B. and Pavic J. (2018). Influence of Aromatherapy on Alleviation of Insomnia Symptoms. *Complementary and Alternative Medicine.* CAM-103. doi: 10.9016/CAM-103/10000103.

[29] Karaca, M., Kara, Ş. M. and Özcan, M. M. (2017). Determination of herb yield and essential oil ratio of some basil (*Ocimum basilicum* L.)

populations. *Ordu University Journal of Science and Technology.* 7(2), 160-169.

[30] Karaca, M. (2017). *Determination of herba yield and essential oil ratio of some reyhan (Ocimum Basilicum L.) populations.* Master Thesis. Ordu University. Institute of Science. Department of Field Crops. 33s. Ordu.

[31] Karik, Ü. (2014). Determination of morphological, yield and quality characteristics of some commercial and local basil (*Ocimum basilicum* L.) Varieties in Menemen ecological conditions. *Journal of Anadolu Aegean Agricultural Research Institute*, 24(2), 10-20.

[32] Kumar, A., Shukla, R., Singh, P., Prakash, B. and Dubey, N.K. (2011). Chemical composition of *Ocimum basilicum* L. essential oil and its efficacy as a preservative against fungal and aflatoxin contamination of dry fruits. *International Journal of Food Science and Technology.* 46(9): 1840–1846.

[33] Makri, O. and Kintzios, S. (2008). *Ocimum* sp. (Basil): Botany, Cultivation, Pharmaceutical Properties, and Biotechnology. *Journal of Herbs Spices and Medicinal Plants.* 13(3): 123-150.

[34] Nacar, Ş. and Tansı, S. (2000). Chemical components of different basil (*Ocimum basilicum* L.) cultivars grown in Mediterranean regions in Turkey. *Israil Journal of Plant Science.* 48(2): 109-112.

[35] Nacar, Ş. (1997). *The effect of different planting densities on yield and quality of basil (Ocimum basilicum L.) plants obtained from different regions.* Cumhuriyet University, Institute of Science (PhD Thesis).

[36] Nguyen, P. M. and Niemeyer, E. D. (2008). Effects of nitrogen fertilization on the phenolic composition and antioxidant properties of basil (*Ocimum basilicum* L.). *Journal of Agricultural and Food Chemistry.* 56(18): 8685–8691. https://doi.org/10.1021/jf801485u PMid:18712879.

[37] Nguyen, P. M., Kwee, E. M. and Niemeyer E. D. (2010). Potassium rate alters the antioxidant capacity and phenolic concentration of basil (*Ocimum basilicum* L.) leaves. *Food Chemistry.* 123(4): 1235–1241.

[38] Nurzyńska-Wierdak, R. (2007). Comparing the growth and flowering of selected basil (*Ocimum basilicum* L.) varieties. *Acta Agrobotanica* 60(2): 127–131.

[39] Okazaki, K., Nakayama, S., Kawazoe K. and Takaishi, Y. (1998). Antiaggregant effects on human platelets of culinary herbs. *Phytotherapy Research.* 12(8): 603–605. https://doi.org/10.1002/(SICI)1099-1573(199812)12:8<603::AID-PTR372>3.0.CO;2-G

[40] Opalchenova, G. and Obreshkova, D. (2003). Comparative studies on the activity of basil--an essential oil from *Ocimum basilicum* L. against multidrug resistant clinical isolates of the genera Staphylococcus, Enterococcus and Pseudomonas by using different test methods. *Journal of Microbiological Methods.* 54(1): 105–110. https://doi.org/10.1016/S0167-7012(03)00012-5.

[41] Özcan, M. and Chalchat, J. C. (2002). Essential Oil Composition of *Ocimum basilicum* and *Ocimum Minimum* in Turkey. *Czech. J. Food. Sci.* 20(6): 223-228.

[42] Özcan, M. M. (2014). *Determination of some yield characteristics and essential oil ratios of selected basil (Ocimum basilicum L.) genotypes according to formation times.* Ordu University Graduate School of Natural and Applied Sciences Department of Field Crops.

[43] Özek, T. Beis, S. H., Demirçakmak. B. and Başer. K. H. C. (1994). Composition of the Essential Oil of *Ocimum basilicum* L. Cultivated in Turkey. *Journal of Essential Oils Research.* 7(2): 203-205.

[44] Pandey, V., Patel, A. and Patra, D. D. (2016). Integrated nutrient regimes ameliorate crop productivity, nutritive value, antioxidant activity and volatiles in basil (*Ocimum basilicum* L.). *Industrial Crops and Products.* 87: 124–131. https://doi.org/10.1016/j.indcrop.2016.04.035.

[45] Purushothaman, B., PrasannaSrinivasan, R., Suganthi, P., Ranganathan, B., Gimbun, J. and Shanmugam, K. (2018). A comprehensive review on *Ocimum basilicum.* *Journal of Natural Remedies.* 18(3), 71-85.

[46] Rai, V., Mani, U. V. and Iyer, U. M. (2009). Effect of *Ocimum sanctum* leaf powder on blood lipoproteins, glycated proteins and

total Amino acids in patients with Non-insulin-dependent Diabetes Mellitus. *Journal of Nutritional & Environmental Medicine.* 7(2): 113–118.

[47] Raina, P., Mundkinajeddu, D., Chandrasekaran, C. V., Agarwal, A., Wagh, N. and Kaul-Ghanekar, R. (2016). Comparative analysis of anti-inflammatory activity of aqueous and methanolic extracts of *Ocimum basilicum* in RAW264.7, SW1353 and human primary chondrocytes. *Journal of Herbal Medicine.* 6(1): 28–36. https://doi.org/10.1016/j.hermed.2016.01.002.

[48] Sestili, P., Ismail, T., Calcabrini, C., Guescini, M., Catanzaro, E., Turrini, E., Layla, A., Akhtar, S. and Fimognari, C. (2018). The potential effects of *Ocimum basilicum* on health: a review of pharmacological and toxicological studies. *Expert Opinion on Drug Metabolism and Toxicology.* 14(7): 679-692.

[49] Siddiqui, B. S., Bhatti, H. A., Begum, S. and Perwaiz, S. (2012). Evaluation of the antimycobacterium activity of the constituents from *Ocimum basilicum* against Mycobacterium tuberculosis. *Journal of Ethnopharmacology.* 144(1): 220–222. https://doi.org/10.1016/j.jep.2012.08.003 PMid:22982011.

[50] Simon, J. E., Morales, M. R., Phippen, W. B., Vieira, R. and Hao, Z. (1999). Basil: A source of aroma compounds and a popular culinary and ornamental herb. *Perspectives on New crops and new uses.* Alexandria: ASHS Press.

[51] Sönmez, Ç., Soysal, A. Ö. Ş., Yildirim, A., Berberoğlu, F. and Bayram, E. (2019). The Effect of Different Form Times on Some Yield and Quality Traits in Green and Purple Basil (*Ocimum basilicum* L.) Types. *Journal of Anatolia Aegean Agricultural Research Institute*, 29(1), 39-49.

[52] Tansı, S. and Nacar, S. (2000). First cultivation trails of lemon basil (*Ocimum basilicum* var. *citriodorum*) in Turkey. *Pakistan Journal of Biological Sciences*, 3(3): 395- 397.

[53] Telci, I. (2005). *Determination of appropriate shape heights in Reyhan (Ocimum basilicum L.) genotypes.*

[54] Telci, I., Bayram, E., Yılmaz, G. and Avcı, A. B. (2006). Variability in essential oil composition of Turkish basils (*Ocimum basilicum* L.). *Biochemical System Ecology* 34(6): 489–497.

[55] Telci, İ., Bayram, E., Yılmaz, G. and Avcı, A. B. (2005). *Local culture in Turkey with basil (Ocimum spp.) genotypes morphological characterization of agronomic and technological characteristics and selection of superior plants* (Final Report). Project No: TOGTAG-3102. TÜBİ TAK.

[56] Tohti, I., Tursun, M., Umar, A., Turdi, S., Imin, H. and Nicholas, M. (2006). Aqueous extracts of *Ocimum basilicum* L. (sweet basil) decrease platelet aggregation induced by ADP and thrombin in vivo arterio-venous shunt thrombosis in vivo. *Thrombosis Research.* 118(6): 733–9. https://doi.org/10.1016/j.thromres.2005.12.011 PMid: 16469363.

[57] Tsai, K. D., Lin, B. R., Perng, D. S., Wei, J. C., Yu, Y. W. and Cherng, J. M. (2011). Immunodulatory effects of aqueous extract of Ocimum basilicum (Linn.) and some of its constituents on human immune cells. *Journal of Medicinal Plants Research.* 5(10): 1873–1883.

[58] Umar, A., Zhou, W., Abdusalam, E., Tursun, A., Reyim, N., Tohti, I. and Moore, N. (2014). Effect of *Ocimum basilicum* L. oncyclo-oxygenase isoforms and prostaglandins involved in thrombosis. *Journal of Ethnopharmacology.* 152(1): 151–155. https://doi.org/10.1016/j.jep.2013.12.051 PMid:24412551.

[59] Uzun, A. (2007). A Research on the *Determination of Some Properties of Reyhan (Ocimum basilicum L.) and Thyme (Origanum vulgare L.) Species that can be used as a drug and spice belonging to the Labiate family.* Master Thesis. Ondokuz Mayıs University. Institute of Science. Department of Field Crops. Samsun. IJCBS. 3:47-52.

In: *Ocimum basilicum*
Editor: Andres A. Walton

ISBN: 978-1-53619-265-0
© 2021 Nova Science Publishers, Inc.

Chapter 4

OCIMUM BASILICUM AS A POTENTIAL ANTI-COVID-19 PLANT: REVIEW ON THE ANTIVIRAL ACTIVITY AND MOLECULAR DOCKING OF SOME OF ITS MOLECULES WITH THE SARS-COV-2 MAIN PROTEASE (MPRO)

Pius T. Mpiana[1,*]*, Etienne M. Ngoy*[1]*, Jason T. Kilembe*[1]*, Carlos N. Kabengele*[1]*, Aristote Matondo*[1]*, Clement L. Inkoto*[2]*, Emmanuel M. Lengbiye*[2]*, Domaine T. Mwanangombo*[1]*, Damien S. T. Tshibangu*[1]*, Koto-te-Nyiwa Ngbolua*[2,3] *and Dorothée D. Tshilanda*[1]

[1]Department of Chemistry, Faculty of Sciences, University of Kinshasa, Democratic Republic of the Congo
[2]Department of Biology, Faculty of Sciences, University of Kinshasa, Democratic Republic of the Congo

[*] Corresponding Author's E-mail: Pius T. Mpiana, ptmpiana@gmail.com.

[3]Department of Basic Sciences, Faculty of Medicine, University of Gbado-Lite, Gbado-Lite, Democratic Republic of the Congo

Abstract

Coronavirus disease, a pandemic that has already caused more than one million deaths worldwide, has not yet found an effective and safe treatment; hence the need to resort to medicinal plants. *Ocimum basilicum* is an edible plant that has also shown several biological properties including antiviral, antisickling, antioxidant, anti-inflammatory, etc. Data collected in the literature have shown that the molecules contained in *O. basilicum* possess antiviral properties against several viruses (Herpes Simplex virus-1 and 2, Human Immunodefiency Virus-1, Adeno virus, Hepatitis B Virus, Enterovirus 71, Coxsackie virus B1, etc.). The molecular docking carried out between some molecules of this plant with the main protease (Mpro) of SARS-CoV-2 involved in the replication of the virus responsible for COVID-19 shows interesting interactions and stables complexes. These molecules could act either alone or in synergy to inhibit viral replication. In addition, aromatic plants including *O. basilicum* are used in Congolese traditional medicine for the treatment of respiratory and inflammatory diseases. The objective of this work is therefore to review the literature on the antiviral activity of *O. basilicum* and to find molecules capable of inhibiting the SARS-CoV-2 main protease. This could permit the use of this plant in the fight against COVID-19 and associated diseases.

Keywords: *Ocimum basilicum*, COVID-19, molecular docking, antiviral activity, Mpro SARS-CoV-2

Introduction

Humanity is facing a new viral disease declared a global pandemic by the World Health Organization (WHO), COVID-19. This disease, which was declared in Wuhan in China in December 2019, is caused by the Severe Acute Respiratory Syndrome of Coronavirus-2 (SARS-CoV-2) which is a seventh virus of the coronavirus family infecting humans (Kumar and Sharma, 2020; Imran et al., 2020; Rane et al., 2020). Currently more than twenty million people are infected, of whom more than one million have died. Patients present symptoms such as fever, cough, aches, diarrhea, respiratory disorders; but some patients are asymptomatic (Al-Jabir et al., 2020; Vellingiri et al., 2020; Tshibangu et al., 2020).

To date, no drug or vaccine has yet been found; on this basis, some drugs such as lopinavir, ritonavir, sofosbuvir, Ribavirin, remdesivir formerly used against SARS-CoV-1, MERS-CoV, HIV and Ebola virus are used for the management of patients with COVID-19 (Cao et al., 2020). On the other hand, chloroquine and hydroxychloroquine have been proposed for the treatment of COVID-19 based on their use as immunomodulators in some coronavirus infections, but their use is the subject of debate among scientists (Liu et al., 2020; Wang et al., 2020; Mpiana et al., 2020).

On the basis of all these problems, traditional medicine appears to be a prime candidate for the treatment of COVID-19 because of its multiple proofs in the treatment of different diseases (Mbadiko et al., 2020, Li-Sheng et al., 2020). Medicinal plants possess several molecules with many biological activities such as antiviral, antioxidant, antihelmintic, antifungal, etc. (Ngbolua et al., 2020). The WHO has revealed that more than 80% of the African population uses these plants to relieve their ailments due to problems related to the accessibility, cost and sometimes toxicity of some modern products (Ngbolua et al., 2016; Tshilanda et al., 2019; Mpiana et al., 2012).

One of the medicinal plants known for its multiple biological properties is *Ocimum basilicum,* belonging to the Lamiaceae family. It is an aromatic plant containing many secondary metabolites, macro and micronutrients and its essential oil contains several chemical compounds such as estragole, linalool, 1,8-cineole, etc., which confers antiviral activities (Tshilanda et al., 2016a).

The aim of this chapter is to review the literature on the antiviral activities of *O. basilicum,* as well as to do a molecular docking on the different molecules that this plant possesses in order to pinpoint the molecule(s) capable of inhibiting the SARS-CoV-2 protease. This will enable us to propose this plant as a candidate in the fight against COVID-19.

MATERIAL AND METHODS

Literature Review

Data published on *O. basilicum,* the chemical compounds (phytochemistry) and their antiviral activities have been collected in the online bibliographic databases, such as: Google scholar, DOAJ, PubMed, PubMed Central, Science Direct, SCIELO, Semantic scholar and Science alert.

Molecular Docking

Protein Preparation

The coronavirus encodes more than one dozen proteins, among these, the 3C-Like protease (3CLpro) is the most studied. The 3CLpro or the COVID-19 virus main protease (Mpro) is a key CoV enzyme which plays a pivotal role in mediating viral replication and transcription, making it an attractive drug target for this virus (Yang et al., 2018).

The crystal structures of Mpro (PDB ID: 6LU7) (Xu et al., 2020) was retrieved from the Protein Data Bank (PDB) and imported into Discovery studio visualizer for identifying the amino acids in the binding pocket. The analyzed structure was further imported into Auto dock 4.2 where the inhibitor and water molecules were removed before the docking and hydrogen atoms were added to the protein in order to correct the ionization and tautomeric states of the amino acid residues. Kollman charges were added to the protein and was saved in pdbqt format. Figure 1 below displays the complex between COVID-19 Mpro and the co-crystallized inhibitor 2GTB (PDB). 2GTB is the main protease found in the CoV associated with the severe acute respiratory syndrome (SARS), and that the main protease in 2019-nCoV shares 96% of similarity with that in SARS (Xu et al., 2020).

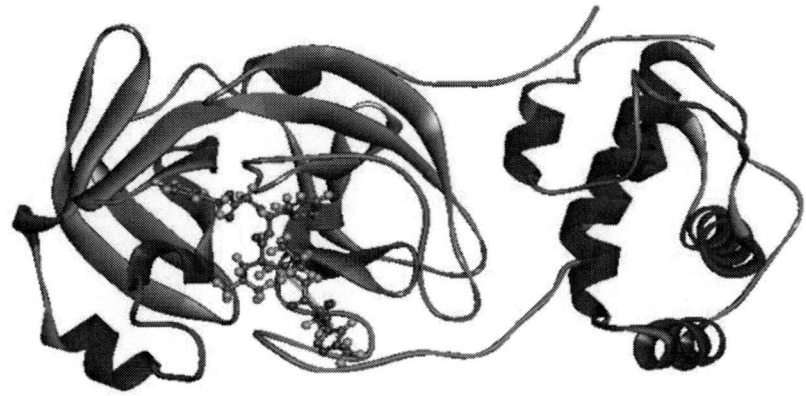

Figure 1. Complex between COVID-19 Mpro and the co-crystallized inhibitor 2GTB (PDB).

Generation of Ligand Dataset

The selected compounds (1-10) used as ligands were drawn using ACD/Marvin Sketch (20.9) (Kim et al., 2017). Figure 2 shows their 2D structures. Furthermore, the ligands were imported into Chem Draw to obtain 3D from 2D. The 3D ligands were saved in .pdf format. The ligands were then imported into Auto dock 4.2 interface tool and saved to pdbqt format.

Docking Strategy

Auto dock tool was used to generate the bioactive binding poses of ligands dataset in the active site of COVID-19 Mpro. The protein coordinates from the bound ligand of 6LU7 was used to define the binding active site. So, scoring function was calculated using the standard protocol of Lamarckian genetic algorithm. The grid map for docking calculations was centered on the target protein. Accelrys Discovery Studio 2019 software was used to model non-bonded polar and hydrophobic contacts in the inhibitor site of 6LU7. The docking result was visualized using Pymol 2.3.4.0 (Figure 4 and 5) and Discovery Studio Visualizer 4.0.

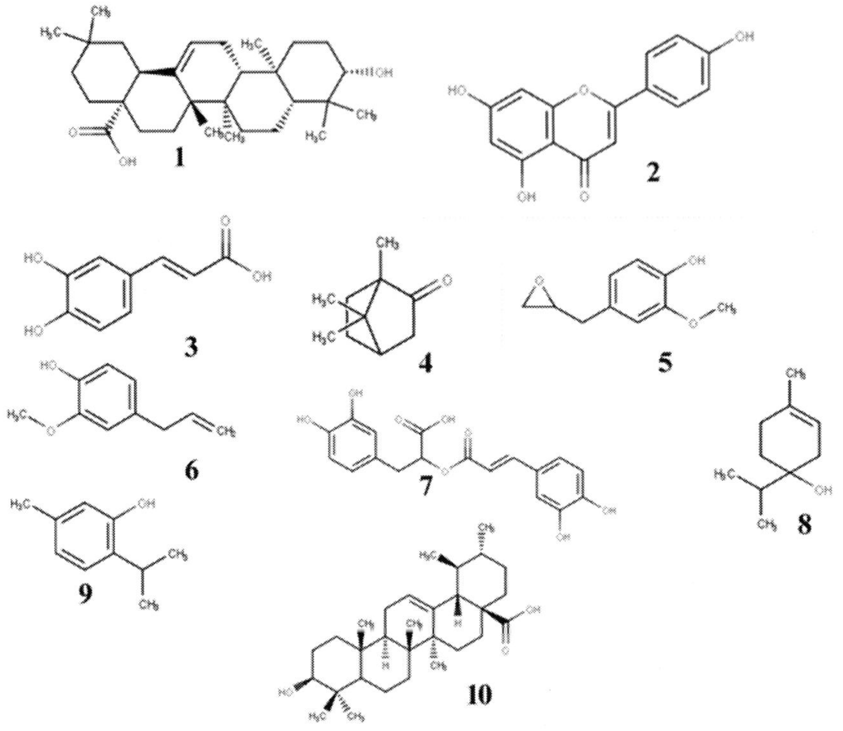

Figure 2. Structures of compounds 1–10 derived from O. basilicum.

RESULTS AND DISCUSSION

Antiviral Activities of *Ocimum Basilicum*

Essential oil and extracts of *O. basilicum* contain many chemical compounds such as estragole, linalool, epi-α-cadinol, α-bergamotene, 1,8-cineol, β-farnescene, α-Guaiene, methylcinnamate, eugenol, eucalyptol, Ursolic acid, rosmarinic acid, oleanolic acid, lithospermic acid, vanillic acid, p-coumaric acid, kampferol, malic acid, tartaric acid, caffeic acid, chicoric acid, caftaric acid, etc. Carbohydrates, proteins, amino acids, thiamin, riboflavin, niacin, zinc, iron, calcium, potassium, magnesium, phosphorus, etc. have also been found in *O. basilicum*. These chemical compounds are indispensable for health and have shown interesting biological effects (Tshilanda et al., 2016a,b; Romeilah et al., 2010; Politeo et al., 2007; Mota et al., 2020; Aburjai et al., 2020; Bobakulov et al., 2020; Piras et al., 2018; Daniel et al., 2011; Bariyah et al., 2012; Pachkore and Dhale 2012; Silva et al., 2008; Javanmardi et al., 2002; Lee and Scagel, 2008).

Table 1 presents some of the major compounds contained in the essential oil as well as in *O. basilicum* extracts and their antiviral activities.

Table 1. Antiviral activity of some chemical compounds of *O. basilicum*

Chemical compounds	Virus	Mode of action	References
Ursolic acid	ADV-8, 11, HSV, HBV, EV, CVB1 HCV,	Inhibition of virus replication	Amber et al., 2017 Chiang et al., 2005 Silva et al., 2008
Linalool	ADV-3,8,11, HBV		
Apigenine	HSV, ADV-3,8,11; HBV; EV, CVB1		
Oleanolic acid	ADV-3, 8, 11		Chiang et al., 2002
Caffeic acid	HSV-1	Inhibits the multiplication	Pandey et al., 2017; Ikeda et al., 2011
	HSV-2, ADV-3		Chiang et al., 2002
Chemical compounds	Virus	Mode of action	References
1,8-cineole	HSV 1 and 2	Destruction of the viral envelope preventing entry of the virus into the host cell.	Kubiça et al., 2014
	IBV	Nucleocapsid (N) protein destruction of the virus	Zoghbi et al., 2007; Yang et al., 2010

Table 1. (Continued)

Chemical compounds	Virus	Mode of action	References
Camphor,	HSV 1 and 2	Destruction of the viral envelope preventing entry of the virus into the host cell.	Kubiça et al., 2014
	IV AH1N1	Inhibition of virus reproduction	Sokolova et al., 2013
Thymol	HSV 1 and 2	Destruction of the viral envelope preventing entry of the virus into the host cell.	Kubiça et al., 2014
Eugenol	HIV-1, HSV-1 and 2	Inhibition of virus replication	Behbahani et al., 2013; Caamal-Herrera et al., 2008
Eugenol epoxide	HIV-1	Inhibition of virus replication	Behbahani et al., 2013
Rosmarinic acid	HSV 1and 2	Inhibition of replication and protease	Chiang et al., 2005
	EV71	inhibits viral IRES	Pandey et al., 2017; Chung et al., 2015
Germacrene	PI-3, RSV	Inhibition of virus replication	But et al., 2009
	HSV-1	Inhibition of virus replication	Venturi et al., 2014
Citral	YFV	Inhibition of virus replication	Gomèz et al., 2013
	HSV-1		Astani et al., 2009
	MNV	Acts directly upon the viral capsid and RNA	Gilling et al., 2014
p-coumaric acid	HSV-2, ADV-11	Inhibition of virus replication	Chiang et al., 2002

Legend: ADV: adenovirus, HSV: herpes simplex virus, HIV: human immunodeficiency virus, YFV: yellow fever virus, MNV: murine norovirus, RSV: respiratory syncytial virus, PI: parainfluenza, EV: enterovirus, IBV: Influenza B virus, CBV: Coxsackie B virus, HBV: hepatitis B virus, HCV: hepatitis C virus, InfV: influenza virus.

From this table, it can be seen that some chemical compounds isolated from the essential oil and extracts of *O. basilicum* such as ursolic acid, rosmarinic acid, caffeic acid, apigenin, estragole, linalool, 1,8-cineole, eugenol, etc., have been found to be present in the essential oil and extracts of *O. basilicum* have shown antiviral activities against several DNA and RNA viruses (HSV, ADV, HIV, EV71, HBV, CVB1, etc.), whose certain are responsible for respiratory and gastrointestinal problems (Amber et al., 2017). With regards to the chemical elements, it was reported that the Zn^{2+} ion inhibits *in vitro* the RNA polymerase activity of coronaviruses and arteriviruses, and the zinc ionophore blocks the replication of these viruses in cell culture (Aartjan te Velthuis et al., 2010).

Molecular Docking Study

Autodock 4.2 was used to assess the binding affinities and main interactions between the COVID-19 Mpro and *O. basilicum* metabolite compounds. Several compounds from *O. basilicum* have been reported to show antiviral bioactivities. Ten compounds from this plant were investigated as potential inhibitors of the COVID-19 Mpro. The free enthalpies obtained from docking of 6LU7 with ten *O. basilicum* compounds are given in Table 2.

The Autodock binding energy values range from -4.79 to -8.55 kcal/mol, and reveal three best docked compounds with highest binding affinities: oleanolic acid/ligand 1 (-8.55 kcal/mol), ursolic acid/ligand 10 (-8.21 kcal/mol) and apigenin/ligand 2 (-7.72 kcal/mol). These three compounds have binding energy higher than that of the ligand reference which has the binding energy computed to -7.40 kcal/mol. Inspection of the table also reveals a strong competition between complexes formed by ligands 3, 4, 5, 6, 7, 8, and 9 with the Mpro. One can easily see that all the complexes formed between these different compounds and the virus protease have 5 kcal/mol has binding energy. Consequently, only molecules 1, 2 and 10 should have better antagonistic properties than the others.

Table 2. Free enthalpies of binding (kcal/mol) from molecular docking calculations

N°	Name	Binding Affinity
1.	Oleanolic acid	-8.55
10.	Ursolic acid	-8.21
2.	Apigenin	-7.72
	Ref_Ligand	-7.40
4.	Camphor	-5.64
8.	Terpinenol	-5.43
7.	Rosmarinic acid	-5.24
9.	Thymol	-5.19
5.	Eugenol epoxyde	-5.12
6.	Eugenol	-5.02
3.	Cafeic acid	-4.79

As depicted in Figure 3, the greatest contribution to the stabilization of the complexes comes from hydrogen bonding interaction that involves O-H and C = O groups of the ligands that can act simultaneously as donor and acceptor (Matondo et al., 2018). Nevertheless, other significant contributions to the stability of the complexes come π-π (stacking and T-shaped) and π-alkyl interactions that are controlled by dispersions forces (Trujillo et al., 2016, Kasende et al., 2017). H-bond distances and angles between the protein target and the 1, 2 and 10 ligands along with the involved groups of ligands in the H-bonds as well as the interacting residues of the three best docked ligands from *O. Basilicum* are gathered in Table 3.

Figure 4 shows the three best docked ligands in the binding pocket of the SARS-CoV-2 main protease.

Table 3. Hydrogen-bonds parameters (distances and angles) derived from docking of COVID-19 Mpro with the three best docked ligands

Ligand	AA Residues	Ligand Group	δ (Å)	θ(°)
1	GLU166	O = C_2	1.39	150
	HIS164	H-O	1.34	133
	THR26	H-O	2.01	164
2	ASP187	H-O	2.01	153
	THR190	H-O	1.87	175
10	GLU166	O = C_2	1.99	185
	GLN189	H-O	2.01	165
	THR24	H-O	2.00	135

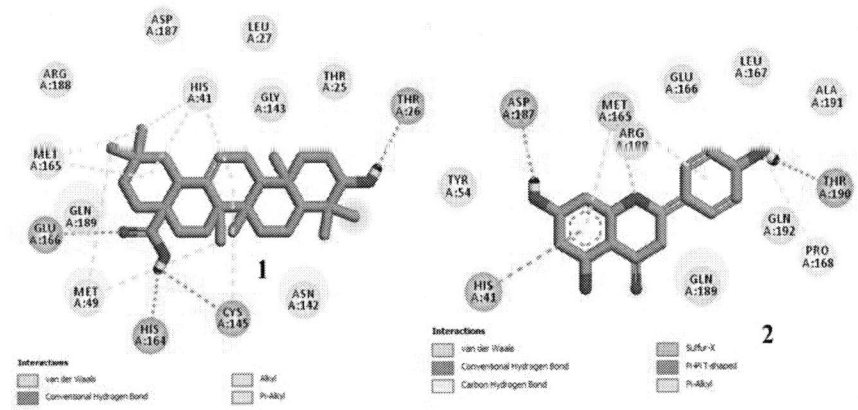

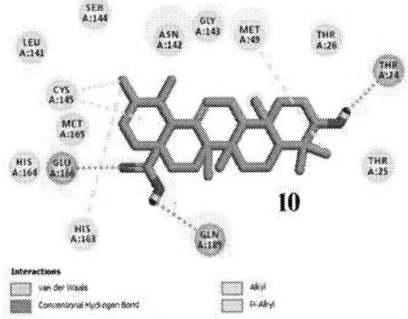

Figure 3. 2D interactions map for ligands 1, 2 and 10 with COVID-19 Mpro.

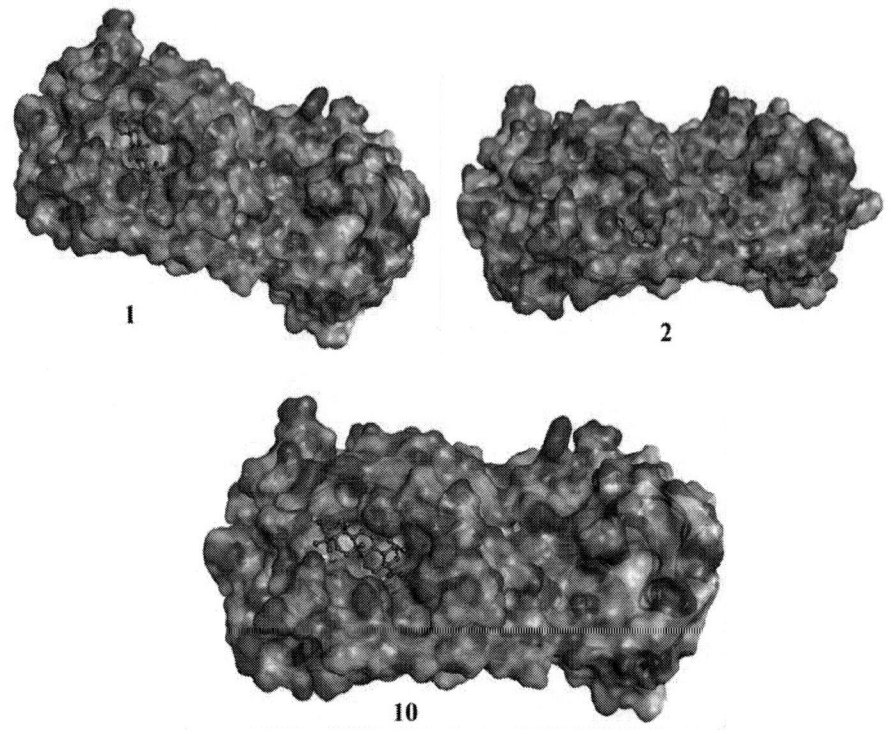

Figure 4. Selected ligand in the binding pocket of COVID-19 Mpro of 6LU7 (1, 2 and 10).

Herbal medicines and purified natural products provide a rich resource for novel antiviral drug development. Proteins such as 3CLpro (Coronavirus main protein), PLpro (papain-like protease), RdRp (RNA-dependent RNA polymerase), S protein (Viral spike glycoprotein), TMPRSS2 (Transmembrane protease serine 2), ACE 2 (Angiotensin converting enzyme 2), AT2 (Angiotensin AT2 receptor) play a key role in COVID-19 viral infection and are considered as potential pharmacological targets (Liu et al., 2020)

Thus, it should be noted that the best anti COVID-19 drug candidate is the molecule that can specifically bind to one of the above-mentioned pharmacological targets to form a stable complex. Thermodynamically, this is a compound with the highest possible binding energy expressed in terms of free enthalpy (ΔG) in absolute terms (Siti et al., 2020). In this study, it was demonstrated by molecular docking that 1, 2 and 10 molecules isolated from the *O. basilicum* plant are potential inhibitors of the protease 3CLPro.

REFERENCES

[1] Aartjan te Velthuis, J. W., Sjoerd van den Worm, H. E., Amy Sims, C., Baric, R.S., Snijder E. J., Van Hemert, M. J. (2010). Zn^{2+} Inhibits Coronavirus and Arterivirus RNA Polymerase Activity In Vitro and Zinc Ionophores Block the Replication of These Viruses in Cell Culture. *Plos Pathogens* 6(XI): 1-10, e1001176. https://doi.org/10.1371/journal.ppat.10 01176.

[2] Aburjai, T. A., Mansi, K., Azzam, H., Dana, A., Alshaer, W., Abuirjei, M. (2020) Chemical compositions and anticancer potential of essential oil from green house cultivated *Ocimum basilicum* leaves. *Indian J. Pharm. Sci.* 82(I):179-184.

[3] Al-Jabir, A., Kerwan, A., Nicola, M., Alsafi, Z., Khan, M., Sohrabi, C., O'Neill, N., Iosifidis, C., Griffin, M., Mathew, G., Agha, R. (2020). Impact of the Coronavirus (COVID-19) pandemic on surgical practice-Part 1 (Review Article). *Science of the Total Environment, Journal Pre-proof.*; DOI : https://doi.org/10.1016/j.ijsu.2020.05.022.

[4] Amber, R., Adnan, M., Tariq, A., Mussarat, S. (2017). A review on antiviral activity of the Himalayan medicinal plants traditionally used to treat bronchitis and related symptoms. *J. of Pharmacy and Pharmacology.* 69:109-122.

[5] Astani, A., Reichling, J., Schnitzler, P. (2009). Comparative Study on the Antiviral Activity of Selected Monoterpenes Derived from Essential Oils. *Phytother. Res*. DOI: 10.1002/ptr.2955.

[6] Bariyah, S., Ahmed, D., Aujla, M. (2012). Comparative Analysis of *Ocimum basilicum* and *Ocimum sanctum*: Extraction Techniques and Urease and alpha-Amylase inhibition. *Pak. J. Chem.* 2(III):134-141.

[7] Behbahani, M., Mohabatkar, H., Soltani, M. (2013). Anti-HIV-1 activities of aerial parts of *Ocimum basilicum* and its parasite *Cuscuta campestris*. *J. Antivir. Antiretrovir.* 5:57-61.

[8] Bobakulov, K., Ozek, G., Ozek, T., Asilbekova, D.T., Abdullaev, N.D., Sagdullaev, S.H. (2020). Essential oils and lipids from the flowers of two varieties of *Ocimum basilicum* L. cultivated in Uzbekistan. *Journal of Essential oil Research* 32(IV):223-230 https://doi.org/10.1080/10412905.2020.1749946.

[9] But, P. P.-H., He, Z.-D., Ma, S.-C., Chan, Y.-M., Shaw, P.-C., Ye, W.-C., Jiang, R.-W. (2009). Antiviral Constituents against Respiratory Viruses from *Mikania micrantha*. *J. Nat. Prod.* 72 (V): 925–928.

[10] Caamal-Herrera, O., Muñoz-Rodríguez, D., Madera-Santana, T., Azamar-Barrios, J.A.(2008). Identification of volatile compounds in essential oil and extracts of *Ocimum micranthum* Willd leaves using GC/MS. *International Journal of Applied Research in Natural Products*. 9(1):31-40.

[11] Cao, B., Wang, Y., Wen, D. (2020). A trial of lopinavir-ritonavir in adults hospitalized with severe COVID-19. *N. Engl. J. Med.*; 10.1056/NEJMoa2001282. 32187464.

[12] Chiang, L. C., Chiang, W., Chang, M. Y., Ng, L. T., Lin, C. C. (2002). Antiviral activity of Plantago major extracts and related compounds in vitro. *Antiviral Research* 55: 53-62.

[13] Chiang, L. C., Ng, L. T., Cheng, P. W., Chiang, W., Lin, C. C. (2005). Antiviral activities of extracts and selected pure constituents of *Ocimum basilicum*. *Clinical and Experimental Pharmacology and Physiology*. 32:811-816.

[14] Chung, Y. C., Hsieh, F. C., Ju Lin, Y., Wu, T. Y., Lin, C. W., Lin, C. T. (2015). Magnesium lithospermate B and rosmarinic acid, two compounds present in *Salvia miltiorrhiza*, have potent antiviral activity against enterovirus infections. *Eur. J. of Pharmacology*. 755:127–133.

[15] Daniel, V. N., Daniang, I. E., Nimyel, N. D. (2011). Phytochemical Analysis and Mineral Elements Composition of *Ocimum Basilicum*

Obtained in JOS Metropolis, Plateau State, Nigeria. *IJET-IJENS.* 11(VI):135-137.

[16] Gilling, D. H., Kitajima, M., Torrey, J. R., Bright, K. R. (2014). Mechanisms of Antiviral Action of Plant Antimicrobials against Murine Norovirus.; *Journal of Environmental Science and Health*, 54 (VII): 608-616. http://dx.doi.org/10.1128/AEM.00402-14.

[17] Gomèz, L. A., Stashenko, E., Ocazionez, R. E. (2014). Comparative Study on *In Vitro* Activities of Citral, Limonene and Essential Oils from *Lippia citriodora* and *L. alba* on Yellow Fever Virus.; *Natural products communications* https://doi.org/10.1177/1934578X1300800 230.

[18] Ikeda, K., Tsujimoto, K., Uozaki, T., Nishide, M., Suzuki, Y., Koyama, A. H. (2011). Inhibition of multiplication of herpes simplex virus by caffeic acid. *Int. J. of Mol Med.*; 28: 595-598.

[19] Imran A and Alharbi O. M. L. (2020). COVID-19: Disease, management, treatment, and social impact. Science of the Total Environment, *Science of The Total Environment* 728 (I): https://doi.org/10.1016/j.scitotenv.2020.138861.

[20] Javanmardi, J., Khalighi, A., Kashi, A., Bais, H. P., Vivanco, J. M. (2002). Chemical Characterization of Basil (*Ocimum basilicum* L.) Found in Local Accessions and Used in Traditional Medicines in Iran. *J. Agric. Food Chem.* 50(XXI):5878–5883.

[21] Kasende, O. E., Matondo, A., Muya, J. T., Scheiner, S. (2017). Interactions between temozolomide and guanine and its S and Se-substituted analogues. *Int. J. Quantum Chem.* 117: 157−169.

[22] Kim, J. H., Yoon, J.-Y., Seo, Y. Y., Choi, S.-K., Kwon, S. J., Cho, I. S., Jeong, M. H., Kim, Y. H., Choi, G. S. (2017). Tyrosinase inhibitory components from *Aloe vera* and their antiviral activity, *J. of Enzyme Inhib. Med. Chem.* 32: 78–83.

[23] Kubiça, T. F., Alves, S. H., Weiblen, R., Lovato, T. L. (2014). In vitro inhibition of the bovine viral diarrhoea virus by the essential oil of *Ocimum basilicum* (basil) and monoterpenes. *Brazilian Journal of Microbiology*.45(I):209-214.

[24] Kumar, D. M. and Sharma, D. (2020). Evaluation of Traditional Ayurvedic Preparation for Prevention and Management of the Novel Coronavirus (SARS-CoV-2) Using Molecular Docking Approach. *ChemRxiv*. Preprint.; https://doi.org/10.26434/chemrxiv.12110214.v2.

[25] Lee, J. and Scagel, C. F. Chicoric acid found in basil (*Ocimum basilicum* L.) leaves. *Food Chemistry* 115(II) 650-656. https://doi.org/10.1016/j.foodchem.2008.12.075.

[26] Li-sheng, W., Yi-ru, W., Da-wei, Y., Qing-quan, L. (2020). A review of the 2019 Novel Coronavirus (Covid-19) based on current evidence. *International Journal of Antimicrobial Agents.*; doi: https://doi.org/10.1016/j.ijantimicag.2020.105948.

[27] Liu, C., Zhou, Q., Li, Y., Garner, L. V., Watkins, S. P, Carter, L. J. (2020). Research and development on therapeutic agents and vaccines for COVID-19 and related human Coronavirus diseases. *ACS Cent Sc*; 25(VI):315-331. doi: https://dx.doi.org/10.1021/acscentsci.0 c00272.

[28] Matondo, A., Mukeba, C. T., Muzomwe, M., Nsimba, B. M., Tsalu, P. V. (2018). Unravelling syn- and anti-orientation in the regioselectivity of carbonyl groups of 5-fluorouracil an anticancer drug toward proton donors. *Chem. Phys. Lett.*; 712: 196–207.

[29] Mbadiko, M. C., Inkoto, L. C., Gbolo, B. Z., Lengbiye, M. E., Kilembe, J. T., Matondo, A., Mwanangombo, D. T., Ngoyi, E. M., Bongo, N. G., Falanga, C. M., Tshibangu, D. S. T., Tshilanda, D. D., Ngbolua, K. N., Mpiana, P. T. (2020). A Mini Review on the Phytochemistry, Toxicology and Antiviral Activity of Some Medically Interesting Zingiberaceae Species. *J. of Complement and Alternative Medic Res.*; 9(IV):44-56. doi.org/10.9734/jocamr/2020/v9i430150.

[30] Mota, I., Sánchez, J. S., Pedro, L. G., Sousa, M. J. (2020). Composition variation of the essential oil from *Ocimum basilicum* L. cv.Genovese Gigante in response to Glomus intraradices and mild water stress at different stages of growth. *Biochemical Systematics and Ecology*. https://doi.org/10.1016/j.bse.2020.104021.

[31] Mpiana, P. T., Ngbolua, K. N., Mudogo, V. (2012).The potential effectiveness of medicinal plants used for the treatment of sickle cell disease in the Democratic Republic of Congo folk medicine: A review. In: Gupta VK, Singh GD (eds.), *Traditional and Folk Herbal Medicine*, Daya Publishing House, New Delhi, India.; 1: 1-11.

[32] Mpiana, P. T., Ngbolua, K. N., Tshibangu, D. S. T, Kilembe, J. T., Gbolo, B. Z., Mwanangombo, D. T. (2020). *Aloe vera* (L.) Burm. F. as a potential anti COVID-19 plant: A minireview of its *antiviral* activity. *Eur. J. of Med. Plants*.; 31(VIII):86-93. doi :10.9734/EJMP/2020/v31i830261.

[33] Ngbolua, K. N., Mbadiko, M. C., Matondo, A., Bongo, N. G., Inkoto, L. C., Gbolo, B. Z., Lengbiye, E. M., Kilembe, J. T., Mwanangombo, D. T., Ngoyi, E. M., Falanga, C. M., Tshibangu, D. S. T., Tshilanda, D. D., Mpiana, P. T. (2020). Review on Ethno-botany, Virucidal Activity, Phytochemistry and Toxicology of Solanum genus: Potential Bio-resources for the Therapeutic Management of Covid-19. *Eur. J. of Nutrition and Food Safety*.;12(7):35-48.doi: 10.9734/EJNFS/2020/v12i730246.

[34] Ngbolua, K. N., Mihigo, S. O., Liyongo, C. I., Ashande, M. C., Tshibangu, D. S. T., Zoawe, B. G., Baholy, R., Fatiany, P.R. & Mpiana, P. T. (2016). Ethno-botanical survey of plant species used in traditional medicine in Kinshasa city (Democratic Republic of the Congo). *Tropical Plant Research*.; 3(II): 413-427.

[35] Pachkore, G. L., Dhale, D. A. (2012). Phytochemicals, vitamins and minerals content of three ocimum species. *Int. J. of Science Innovations and Discoveries*. 2(I): 201-207.

[36] Pandey, S, Singh S. K., Kumar N., Manjhi R. 2017. Antiviral, antiprotozoal, antimalarial and insecticidal activities of *Ocimum gratissimum* L. *Asian Journal of Pharmaceutical Research and Development*. 5(V):1-9.

[37] Piras, A., Gonçalves, M. J., Alves, J., Falconieri, D., Porcedda, S., Maxia, A. (2018). *Ocimum tenuiflorum* L. and *Ocimum basilicum* L., two spices of Lamiaceae family with bioactive essential oils.

*Industrial crops and Products.*113: 87-89. https://doi.org/10.1016/j.indcrop.2018.01.024.

[38] Politeo, O., Jukic, M., Milos, M. (2007). Chemical composition and antioxidant capacity of free volatile aglycones from basil (*Ocimum basilicum* L.) compared with its essential oil. *Food Chemistry.*;101:379-385, doi:10.1016/j.foodchem.2006.01.04.

[39] Rane, J. S., Chatterjee, A., Kumar, A., Ray, S. (2020). Targeting SARS-CoV-2 Spike Protein of COVID-19 with Naturally Occurring Phytochemicals: An *in Silco* Study for Drug Development. *Chem. Rxiv.*, Preprint.; https://doi.org/10.26434/chemrxiv.12094203.v1

[40] Romeilah, R. M., Fayed, S. A., Mahmoud, G. I. (2010). Chemical Compositions, Antiviral and Antioxidant Activities of Seven Essential Oils. *J. of Applied Sciences Research.* 6(I):50-62.

[41] Silva, M. G. V., Vieira, I. G. P., Mendes, F. N. P, Albuquerque, I. L., dos Santos, R. N., Silva, F. O. (2008). Variation of Ursolic Acid Content in Eight *Ocimum* Species from Northeastern Brazil. *Molecules.*; 13:2482-2487, DOI: 10.3390/molecules13102482.

[42] Siti, K., Hendra, K., Rizki, K., Suhartati, S., Soetjipto, S. (2020). Potential inhibitor of Covid-19 Main Protease (Mpro) from several medicinal plant compounds by molecular docking study. *Preprints* 2020030226 doi: 10.20944/preprints202003.0226.v1.

[43] Sokolova, A. S., Yarovaya, O. I., Shernyukov, A. V., Pokrovsky, M. A., Pokrovsky, A. G., Lavrinenko, V. A., Zarubaev, V. V., Tretiak, T. S., Anfimov, P. M., Kiselev, O. I., Beklemishev, A. B., Salakhutdinov, N. F. (2013). New quaternary ammonium camphor derivatives and their antiviral activity, genotoxic effects and cytotoxicity. *Bioorganic and medicinal chemistry* 21 (XXI) 6690-6698. https://doi.org/ 10.1016/j.bmc.2013.08.014.

[44] Trujillo, C. and Sanchez-Sanz, G. (2016). A Study of π–π Stacking Interactions and Aromaticity in Polycyclic Aromatic Hydrocarbon/ Nucleobase Complexes, *Chem. Phys. Chem.*; 17:395-405.

[45] Tshibangu, D. S. T, Matondo, A., Lengbiye, E. M., Inkoto, C. L., Ngoyi, E. M., Kabengele, C. N., Bongo, G. N., Gbolo, B. Z., Kilembe, J. T., Mwanangombo, D. T., Mbadiko, C. M., Mihigo, S.

O., Tshilanda, D. D., Ngbolua; K. N. & Mpiana, P. T. (2020). Possible Effect of Aromatic Plants and Essential Oils against COVID-19: Review of Their Antiviral Activity. *Journal of Complementary* and *Alternative Medical Research.*; *11*(I): 10-22. https://doi.org/10.9734/jocamr/ 2020/v11i130175.

[46] Tshilanda, D. D., Babady, P. B., Onyamboko, D. N. V., Tshiongo, C. M. T., Tshibangu, D. S. T., Ngbolua, K. N. (2016). Chemo-type of essential oil of *Ocimum basilicum* L. from DR Congo and relative in vitro antioxidant potential to the polarity of crude extracts. *Asian Pac J. Trop. Biomed.* 6(XII): 1022-1028.

[47] Tshilanda, D. D., Mutwale, P. K., Onyamboko, D. V. N., Babady, P. B., Tsalu, P. V., Tshibangu, D. S. T. (2016). Chemical Fingerprint and Anti-Sickling Activity of Rosmarinic Acid and Methanolic Extracts from Three Species of Ocimum from DR Congo. *Journal of Biosciences and Medicines* 4:59-68.

[48] Tshilanda, D. D., Inkoto, C. L., Mpongu, K., Mata, S., Mutwale, P. K., Tshibangu, D. S. T., Bongo, G. N., Ngbolua, K. N., Mpiana, P. T. (2019). Microscopic Studies, Phytochemical and Biological Screenings of *Ocimum canum*. *International Journal of Pharmacy and Chemistry.* 5(V): 61-67, doi: 10.11648/j.ijpc.20190505.13.

[49] Vellingiri, B., Jayaramayya, K., Iyer, M., Narayanasamy, A., Govindasamy, V., Giridharan, B., Ganesan, S., Venugopal, A., Venkatesan, D., Ganesan, H., Rajagopalan, K., Rahman, P.K.S.M., Cho, S.-G., Kumar, N.S., Subramaniam, M.D. (2020). COVID-19: A promising cure for the global panic. *Science of the Total Environment.*; https://doi.org/10.1016/j.scitotenv.2020.138277.

[50] Venturi, C. R., Danielli, L. J., Klein, F., Apel, M. A., Montanha, J. A., Bordignon, S. A. L., Roehe, P. M., Fuentefria, A. M., Henriques, A. T. (2014). Chemical analysis and in vitro antiviral and antifungal activities of essential oils from *Glechon spathulata* and *Glechon marifolia.*; *Journal of pharmaceutical biology* 53(V):682-688. https://doi.org/10.3109/13880209.2014.936944.

[51] Wang, M., Cao, R., Zhang, L., Yang, X., Liu, J., Xu, M. (2020). Remdesivir and chloroquine effectively inhibit the recently *emerged*

novel coronavirus (2019-nCoV) in vitro. *Cell Res.* https://doi.org/10.1038/s41422–020– 0282–0.

[52] Xu, Z., Peng, C., Shi, Y., Zhu, Z., Mu, K., Wang, X. (2020). *Nelfinavir was predicted to be a potential inhibitor of 2019-nCoV main protease by an integrative approach combining homology modelling, molecular docking and binding free energy calculation.* Biorxiv doi: https://doi.org/10.1101/2020.01.27.921627.

[53] Yang, L., Wen, K. S., Ruan, X., Zhao, Y. X., Wei, F., Wang, Q. (2018). Response of plant secondary metabolites to environmental factors, *Molecules*. 23: 1–26.

[54] Yang, Z., Wu, N., Fu, Y., Yang, G., Wang, W., Zu, Y. (2010). Anti-Infectious Bronchitis Virus (IBV) Activity of 1,8-cineole: Effect on Nucleocapsid (N) Protein. *J. of Biomolecular Structure and Dynamics*. 28(III):323-330.

[55] Zoghbi, M. G. B., Oliveira, J., Andrade, E. H. A., Trigo, J. R., Fonseca, R. C. M., Rocha, A. E. S. (2007). Variation in Volatiles of *Ocimum campechianum* Mill. and *Ocimum gratissimum* L. Cultivated in the North of Brazil. *Journal of Essential Oil Bearing Plants* 10 (III): 229-240. https://doi.org/10.1080/0972060X.2007.10643547.

BIOGRAPHICAL SKETCH

Pius T. Mpiana

Affiliation: University of Kinshasa, faculty of science, department of chemistry

Education: PhD, 2003 in Chemistry,

E-mail: ptmpiana@gmail.com; pt.mpiana@unikin.ac.cd

Websites:

https://sites.google.com/site/profpiusmpiana/home/cv-du-prof-pius-mpiana
http://scholar.google.com/citations?user=J1m2zdkAAAAJ&hl=en
http://www.researchgate.net/profile/Pius_Tshimankinda_MPIANA/

Research and Professional Experience:

Research domain: medicinal chemistry, environmental chemistry and biophysics of biomacromolecules.

Scientific organization membership: Member of some scientific organization inluding the "Société française d'ethnopharmacologie"

Editorial board membership: Member of editorial board of many scientific journal including:

- Journal of computational methods in molecular design;
- Asian pacific journal of tropical biomedicine;
- Journal of coastal life medicine;
- -Research Journal of medicinal plants;
- International journal of biological chemistry

Seminar Participation: Participation in many scientific seminar and workshop including:

- College on Medical Physics; ICTP, Trieste, Italy (4th to 29th September 2006);
- Natural Product Research in East and Central Africa NAPRECA, 13th Symposium, Kinshasa, DRC (10th to 14th August, 2009);

- Stakeholders Conference S&T cooperation between Europe and Africa: status and way forward, Mombasa, Kenya (10th to 11th November, 2009);
- College on medical Physics; ICTP, Trieste, Italy (13th September to 01 October, 2010);
- Conference on Molecular Aspects of Cell Biology: A Perspective from Computational Physics, Trieste, Italy (11th to 15th October, 2010);
- Colloque Panafricain-Paneuropeen sur "Chimie et Ressources Naturelles," Cotonou, Benin (13th to 16th April, 2015);
- Chemical security and safety in Central and East Africa, CoE CBRN-AIEA, Brunswick, Germany (20.-24. November 2017)

Research Projects directed or co-directed: Responsible of research project WHO/TDR/970015 (1998-1999);

- Responsible of research project WHO/TDR/980594 (1999-2001);
- Co-responsible of research project TWAS Grant N° 04-025 LDC/CHE/AF/AC (2004-2006);
- Co-responsible of research project TWAS Grant N° 07-077 LDC/CHE/AF/AC;
- Co-responsible of research project TWAS Grant N° 08-030/LDC/CHE/AF/AC-UNESCO FR-3 240 204448;
- Supervisor of IFS Research Grant F/4921-1;
- Responsible of project ARES (2015); -Responsible of project TWAS Grant N°15-156 RG/CHE/AF/AC_G – FR3240287018 (2016-2017)

Recent master and Doctoral Thesis directed or co-directed: Director of four doctoral thesis and more than 10 master theses

Professional Appointments:

Current positions:

- Vice head of chemistry department in charge of research, science faculty, University of Kinshasa, DRC (from 2010)
- Professor (Ordinary Professor) at science faculty, chemistry department, University of Kinshasa, DRC.
- Visitor Professor at some Congolese Universities (Kisangani, Bukavu, Kongo …)
- Expert at DR Congo Nuclear Regulatory Authority (CNPRI)
- Expert and member of national team of CoE CBRN
- Member of national scientific team for the management of COVID-19

Previous positions:

- Senior associate, The Abdus Salam Centre of Theoretical Physics ICTP, Trieste, Italy (2009-2015)
- Full Professor at Science Faculty, Chemistry department, University of Kinshasa, DRC (2010-2016)
- Associate professor at Science faculty, Chemistry department, University of Kinshasa, DRC (2003-2010)
- Head of research division at national programme for valorisation of traditional medicine, Congolese Ministry of public health (2002-2005).
- Senior Research and Teaching assistant, University of Kinshasa,DRC (19994-2003)
- Junior Research and teaching Assistant, University of Kinshasa, RDC (1988-1994)

Honors:

Prizes:

- Diploma and medal of scientific merit of DR Congo, 2010
- "Diplôme d'honneur et de mérite du meilleur professeur d'université congolaise pour l'année 2008;"
- "Prix africain pour la liberté et le développement 2009"
- "Prix congolais de mérite civique et patriotique pour l'année 2008-2009"
- "Diplôme d'honneur et d'excellence de la république démocratique du Congo pour l'année 2007-2008"

Patents:

- Shode, F.O.; Koorbanally, N.; Mpiana, P.T; Tshibangu, D.S.T.; Oyedeji, O.O.; Habila, J.D.; University of kwazulu natal, *In vitro* Anti-sickling activity of Betulinic Acid, Oleanolic acid and their derivatives, Wolrd International Property Organisation WIPO, patent N° WO2011/064710 A1. Available on: http://www.google.com/patents/WO2011064710A1?cl=en
- US. Patent Apr. 1, 2014 US 8,685,469 B2 http://www.google.com/patents/US8685469
- South African Patent A & A Ref: P47863ZP00

Publications from the Last 3 Years:

1) Tshibangu, D. S. T; Matondo, A.; Lengbiye, E. M.; Inkoto, C. L.; Ngoyi, E. M.; Kabengele, C. N.; Bongo, G. N.; Gbolo, B. Z.; Kilembe, J. K.; Mwanangombo, D. T.; Mbadiko, C. M.; Shetonde, O. M.; Tshilanda, D. D.; Ngbolua, K. N. and Mpiana, P. T. 2020. Possible Effect of Aromatic Plants and Essential Oils against COVID-19:

Review of Their Antiviral Activity Journal of Complementary and Alternative Medical Research 11(1): 10-22.

2) Kitadi, J. M.; Inkoto, C. L.; Lengbiye, E. M.; Tshibangu, D. S. T.; Tshilanda, D. D.; Ngbolua, K. N.; Taba, K. M.; Mbala, B. M.; Schmitz, B. and Mpiana, P.T. 2020. Mineral Content and Antisickling Activity of *Annona senegalensis, Alchornea cordifolia* and *Vigna unguiculata* Used in the Management of Sickle Cell Disease in the Kwilu Province (Congo, DR) *International Blood Research & Reviews 11(3): 18-27* doi: 10.9734/IBRR/2020/v11i330131.

3) Kwembe, J. T. K.; Onautshu, D. O.; Mpiana, P. T.; Bekaert, B. and Haesaert, G. 2020. Antifungal activity on *Lasiodiplodia theobromae* and phytochemical study of *Mitracarpus villosus* and *Moringa oleifera* from kisangani (d.r.congo) *European journal of pharmaceutical And medical research* 7(10): 125-133.

4) Mpiana, P. T.; Ngbolua, K. N.; Tshibangu, D. S. T.; Kilembe, J. T.; Gbolo, B. Z.; Mwanangombo, D. T.; Inkoto, C. L; Lengbiye, E. M.; Mbadiko, C. M.; Matondo, A.; Bongo, G. N.; Tshilanda, D. D. 2020. Identification of potential inhibitors of SARS-CoV-2 main protease from *Aloe vera* compounds: A molecular docking study *Chemical Physics Letters* 754 (2020) 137751 https://doi.org/10.1016/j.cplett.2020.137751.

5) Mpiana, P. T.; Ngbolua, K. N.; Tshibangu, D. S. T; Kilembe, J. T; Gbolo, B. Z.; Mwanangombo, D. T.; Inkoto, C. L., Lengbiye, E. M.; Mbadiko, C. M.; Matondo, A.; Bongo, G. N.; Tshilanda, D. D. 2020. *Aloe vera* (L.) Burm. F. as a Potential Anti-COVID-19 Plant: A Mini-review of Its Antiviral Activity *European Journal of Medicinal Plants 31(8): 86-93* doi: 10.9734/EJMP/2020/ v31i830261.

6) Ngobua, K. N.; Mbadiko, C. M.; Matondo, A.; Bongo, G.N.; Inkoto, C. L.; Gbolo, B. Z.;Lengbiye, E. M.; Kilembe, J. T.; Mwanangombo, D.T.; Ngoyi, E.M.; Falanga, C.M.; Tshibangu, D.S.T.; Tshilanda, D.D. and Mpiana, P.T. 2020. Review on Ethno-botany, Virucidal Activity, Phytochemistry and Toxicology of *Solanum* genus: Potential Bio-resources for the Therapeutic Management of Covid-19. *European*

Journal of Nutrition & Food Safety 12(7): 35-48 doi: 10.9734/ EJNFS/2020/v12i730246.

7) Mbadiko, C. M.; Inkoto, C. L.; Gbolo, B. Z.; Lengbiye, E. L.; Kilembe, J. T.; Matondo, A.; Mwanangombo, D. T; Ngoyi, E. M.; Bongo, G. N.; Falanga, C. M.; Tshibangu, D. S. T; Tshilanda, D. D. T.; Ngbolua, K. N. and Mpiana, P. T. 2020. A Mini Review on the Phytochemistry, Toxicology and Antiviral Activity of Some Medically Interesting Zingiberaceae. *Species Journal of Complementary and Alternative Medical Research*, 9(4): 44-56 doi:10.9734/JOCAMR/ 2020/ v9i430150.

8) Nsimba, B. M.; Basosila, N. L.; Kayembe, J-C. K.; Mbuyi, D. M.; Matondo, A.;. Bongo, G. N.; Ngbolua, K. N. and Mpiana, P. T. 2020. Semi-empirical Approach on the Methanogenic Toxicity of Aromatic Compounds on the Biogas Production. *Asian Journal of Applied Chemistry Research 5(4): 34-50* doi: 10.9734/AJACR/2020/v5i4 30146.

9) Kamienge, M. K.; Toke, N. N.; Mayanu, B. P.; Assi, R. L.; Gbatea, A. K.; Masengo, C. A.; Mpiana, P. T. and Ngbolua, K. N. 2020. Pouvoir germinatif, croissance et capacité de séquestration de carbone de *Pterygota bequaertii* De Wild (Malvaceae) dans les conditions éco-climatiques de Kinshasa en République Démocratique du Congo. *International Journal of Applied Research* 6(6): 74-80. [Germination power, growth and carbon sequestration capacity of Pterygota bequaertii De Wild (Malvaceae) in the eco-climatic conditions of Kinshasa in the Democratic Republic of Congo.]

10) Ashande, C. M; Djolu, R. D.; Ngambika, G. K.; Aundagba, J. M. P; Amisi, C. M.; Baholy, R. R.; Mpiana, P. T.; Mudogo, V. and Ngbolua, K. N. 2020. Profil épidémiologique et clinique du Paludisme et de la drépanocytose à l'hôpital général de référence de gbado-lite (Nord-Ubangi) en république démocratique du Congo. *International Journal of Applied Research* 6(2): 240-246. [Epidemiological and clinical profile of Malaria and sickle cell disease at the general referral hospital of gbado-lite (Nord-Ubangi) in the Democratic Republic of Congo]

11) Liyongo Inkoto, C. L.; Jean-Pierre Kayembe, P. J. K.; Mpiana, P. T.; Ngbolua, K. N. 2020.A review on the Phytochemistry and Pharmacological properties of *Picralima nitida* Durand and H. (Apocynaceae family): A potential antiCovid-19 medicinal plant species. *Emer. Life Sci. Res.* (2020) 6(1): 64-75 https://doi.org/10.31783/elsr.2020.616475.
12) Kwembe, J. T. K.; Mbula, J. P., Onautshu, O.; Mpiana, P. T.; Haesaert, G. 2020. *In vitro* evaluation of antifungal activity of *Aloe vera, Moringa oleifera* and *Newbouldia laevis* on the Strain of *Lasiodiplodia theobromae* in Region of Kisangani / DR CONGO *Scholars Bulletin.* 6(5): 111-118 doi:10.36348/sb.2020.v06i05. 002.
13) Modeawi, M. N.; Djolu, R. D.; Masengo, C. A.; Falanga, C. M.; Lengbiye, E. M; Inkoto, C. L.; Gbolo, B. Z.; Ridwan, M.; Mpiana, P. T.; Mudogo, V.; Ngbolua, K. N. 2020. Congolese Medicinal Plant biodiversity as Source of AntiCOVID-19 inhibitors: An economic good in the light of Comparative Advantages Theory of Recardo. *Budapest International Research in Exact Sciences* (BirEx) 2(3) 298-309.
14) Suami, R. B; Sivalingam, P., Al Salah, D. M.; Dominique Grandjean, D.; Mulaji, C. K., Mpiana, P. T.; Breider, F.; Otamonga, J-P.; Poté, J. 2020. Heavy metals and persistent organic pollutants contamination in river, estuary, and marine sediments from Atlantic Coast of Democratic Republic of the Congo, *Environmental Science and Pollution Research* https://doi.org/10. 1007/s11356-020-08179-4.
15) BomoiMatita, H. J, Bongo, G. N.; Mpiana P. T., Ukondalemba, L.; Mindele; Nzau, M.; Lobota, L; Booto, B., Mbadiko, C. M. 2020. A mini-review on Pollution of water resources: causes and consequences *GSJ* 8 (3): Online.
16) Masunda, A. T.; Inkoto, C. L.; Masengo, C. A; Bongili, S. B.; Kanza, J-P. B.; Legbiye, E. M.; Ngbolua, K. N.; Mpiana, P. T. 2020. Traditional uses, Physical properties, Phytochemistry and Bioactivity of *Lippia multiflora* Moldenke (Verbenaceae): *A Mini-review Discovery Phytomedicine*, 7(1): 19-26.

17) Bishola, T. T.; Ngbolua, K. N.; Masengo, C. A; Gbolo, B. Z.; Bongo, G. N.; Tshibangu, D. S. T; Tshilanda, D. D.; Mudogo, V.; Mpiana, P.T. 2020. Antibacterial Activity of Three Validated Antisickling Plant Species from the Great Apes Pharmacopoeia in Congo-Kinshasa. *International Journal of Biomedical and Clinical Sciences*, 5(1): 33-40.

18) Oleko wa Oleko, J. D. and Mpiana, P.T. 2020. Oppositions to Vaccination Activities against Poliomyelitis in the province of Tshopo R.D.Congo, *J. of Advancement in Medical and Life Sciences.* 7(3):05. doi: 10.5281/zenodo.3754547.

19) Mukeba, B.; Ngbolua, K. N.; Ngombe, K.; Mpiana, P. T.; Mukoko, B., Mutwale, K.; Kabena, N.; Ngondo, M.; Mbemba, F. 2019. Review on ethnobotany, phytochemitry and bioactivity of the Tropical medicinal plant species *Harungana madagascariensis* Lam. ex Poiret. (*Hypericaceae*) *Discovery Phytomedicine* 7(3): 138-144.

20) Ngbolua, K N ; Inkoto, C. L.; Mongo, N. L.; Ashande, C. M.; Masens, V. B.; Mpiana, P. T. 2019. Étude ethnobotanique et floristique de quelques plantes médicinales commercialisées à Kinshasa, République Démocratique du Congo. *Rev. Mar. Sci. Agron. Vét.* 7 (1): 118-128. [Ethnobotanical and floristic study of some medicinal plants marketed in Kinshasa, Democratic Republic of Congo]

21) Ngbolua, K. N.; Inkoto, C. L.; Mongo, N. L.; Ashande, C. M.; Masens, Y. B.; Mpiana, P. T. 2019. Étude ethnobotanique et floristique de quelques plantes médicinales commercialisées à Kinshasa, République Démocratique du Congo. *Rev. Mar. Sci. Agron. Vét.* (2019) 7 (1): 118-128. [Ethnobotanical and floristic study of some medicinal plants marketed in Kinshasa, Democratic Republic of Congo.]

22) Kafuti, Y. S.; Ojerinde, O. S; Balogun, O.; Alemika, T. E.; Taba, K. M; Mpiana, P. T.; Kindombe, N. M. 2019. Antioxidant and Antiproliferative Activities of the Stem Bark Extract and Fractions of *Boswellia dalzielii* Hutch. *International Journal of Pharmacognosy and Phytochemical Research* 11(3):177-182.

23) Suami, R. B.; Al Salah, D. M. M.; Kabala, C. D.; Otamonga, J.-P.; Mulaji, C.K.; Mpiana, P. T.; John W. Pote, J.W. 2019. *Assessment of*

metal concentrations in oysters and shrimp from Atlantic Coast of the Democratic Republic of the Congo Heliyon 5 (2019) e03049 1-9 https://doi.org/10.1016/j.heliyon.2019.e03049.

24) Kitadi, J. M.; Inkoto, C. L.; Lengbiye, E. M.; Tshibangu, D. S. T; Tshilanda, D. D.; Ngbolua, K. N.; Taba, K. M. Mbala, B. M; Schmitz, B.; Mpiana, P. T. 2019. Antisickling activity and mineral content of *Hura crepitans, Alternanthera bettzickiana* and *Dissotis brazzae*, plants used in the management of sickle cell disease in kwilu province Democratic Republic of the Congo, 6(12).

25) Kitadi, J. M.; Inkoto, C. L.; Lengbiye, E. M.; Taba, K.; Mbala, M.; Tshibangu, D. S. T; Tshilanda, D. D.; Ngbolua, K. N.; Mpiana, P. T. 2019. Nutritional value, phytochemistry and pharmacological activities of *Vigna unguiculata* (l.) Walp.: a review. *European journal of pharmaceutical And medical research*,6(10): 94-100.

26) Bomoi, M. H. J; Booto, B. E.; Bamvingana, K. C.; Kemfine, L. L.; Ngoy, K. S. A.; Lobota L. J., Ngando, B. T.; Nzau, M. L.; Mulumba, K. P. et Mpiana P. T. 2019. Etude des instruments juridiques de la RD Congo sur la gestion des produits chimiques, *International Journal of Innovation and Scientific Research* 42(2): 94-103.

27) Oleko WA Oleko, J-D.; Mpiana, P. T. 2019. Oppositions to vaccination activities against polio in the province of TSHOPO / R.D. Congo; *International Journal of Medical Science Clinical Research* 1(1): 09-13.

28) Muya, J. T.; Mwanangombo, D. T; Tsalu, P. V.; Mpiana, P. T.; Tshibangu, D.S.T; Chung, H. 2019. *Conceptual DFT study of the chemical reactivity of four natural products with anti-sickling activity.* Spinger Nature applied Sciences 1:1457 https://doi.org/10. 1007/s4245 2-019-1438-8.

29) Tshilanda, D. D.; Inkoto, C. L.; Mpongu, K.; Mata, S.; Mutwale, P. K.; Tshibangu, D. S. T.; Bongo, G. N.; Ngbolua, K. N.; Mpiana, P. T. 2019. Microscopic Studies, Phytochemical and Biological Screenings of *Ocimum canum. International Journal of Pharmacy and Chemistry* 5(5): 61-67 doi: 10.11648/j.ijpc.20190505.13.

30) Kapembo, M. L.; Al Salah, D. M. M.; Thevenon, F.; Laffite, A.; Bokolo, M. K.; Mulaji, C. K.; Mpiana, P. T.; Poté J. 2019. Prevalence of water-related diseases and groundwater (drinking-water) contamination in the suburban municipality of Mont Ngafula, Kinshasa (Democratic Republic of the Congo), *Journal of Environmental Science and Health*, Part A, doi: 10.1080/ 10934529.2019.1596702.
31) Oleko WA Oleko, J. D.; Tongo R. N.; Lohohola, P. O; Likoko, J. K.; Chantal N Zingabako, C. N.; Mpiana, P. T., 2019. What epidemiological profile of breast cancer patients treated at university clinics in Kisangani, DR Congo?. *International Journal of Medical and Health Research* 5(8): 35-39.
32) Oleko Wa Oleko, J. D., Tshibanda, J. M.; Tongo, N. R. and Mpiana, P.T. 2019. Evaluation of nutritional status of children in nursery schools in kisangani city, dr congo. *European Journal of Pharmaceutical and Medical Research* 6(9) 59-62.
33) Gbolo, Z. B.; Ngbolua, K. N.; Mpiana, P. T.; Ndanga, B.A.; Pangodi, A. J-M.; Masengo, A. C.;Mudogo, V. 2019. Evaluation of the Clinical Efficiency of an Antisickling Polyherbal Formula Drepanoalpha in a Sickle cell disease Patient in Gbado-Lite City (Democratic Republic of the Congo) by Quantum Magnetic Resonance Analyzer. *Britain International of Exact Sciences (BIoEx) Journal*, 1 (1): 36-48.
34) Kitadi, J. M.; Lengbiye, E. M.; Gbolo, B. Z.; Inkoto, C. L.; Muanyishay, C. L.; Lufuluabo, G. L.; Tshibangu, D. S. T; Tshilanda, D. D.; Mbala, B. M.; Ngbolua, K. N.; Mpiana, P. T. 2019. *Justicia secunda* Vahl species: Phytochemistry, Pharmacology and Future Directions: *A mini-review*. Discovery Phytomedicine 6 (4): 157-171. doi:10.15562/ phytomedicine.2019.93.
35) Lengbiye, M. E., Ngbolua, K. N.; Messi, L. M.; Mbembo wa Mbembo, B.; Bongo, G. N., Mutwale, K. P., Ngombe, K. N., Ngo Mbing, J.; Pegnyemb, D.E.; Mpiana, P. T. 2019. In vitro Evaluation of the Anti-scavenging and Anthelmintic Activities of *Artocarpus heterophyllus* LAM Leaves (Moraceae) in the Democratic Republic of Congo. *International Journal of Biomedical Engineering and Clinical Science* 2019; 5(2): 14-22 doi: 10.11648/j.ijbecs.20190502.11.

36) Ngbolua, K. N.; Ngiala, G. N.; Inkoto, C. L.; Masengo, C. A.; Lufuluabo, G. L.; Janvier Mukiza, J. Mpiana, P. T., 2019. A mini-review on the Phytochemistry and Pharmacology of the medicinal plant species *Persea americana* Mill. (Lauraceae) *Discovery Phytomedicine*, 6(3): 102-111. doi:10.15562/phytomedicine. 2019.99.
37) Ngbolua, K. N.; Ngambika, G. K.; Amisi, C. M.; Ashande, C. M.; Gbolo, B. Z.; Bongo, G. N.; Djolu, R. D.; Mukiza, J.; Mpiana, P. T. 2019. Evaluation of Anti-malarial Drugs Use in Gbadolite Health Area in Democratic Republic of the Congo: A Cross-Sectional Study. *Bioscience and Bioengineering* 5 (1): 1-10.
38) Ngunde-te-Ngunde, S.; Lengbiye, E. L.; Tshidibi, D. J.; Kengo, V. F.; Djolu, R. D.; Masengo, C. A.; Gbolo, B. Z.; Iteku, B. J.; Mpiana, P. T.; Ngbolua, K. N. 2019. Antisickling and Antibacterial Activities of *Anthocleista schweinfurthii Gilg.* (Gentianaceae) from Non-human Primates Pharmacopoeia in Democratic Republic of the Congo. *Budapest International Research in Exact Sciences (BirEx) Journal*, 1(3): 14-20 doi: *https://doi.org/10.33258/ birex.v1i3.345*.
39) Mbadiko, C. M.; Bongo, G. N.; Mindele, L. U.; Ngbolua, K. N.; Mpiana, P. T.; Ngombe, N. K.; Mutwale, P. K.; Mbemba, T. F. 2019. Effect of Drying on the Composition of Secondary Metabolites in Extracts from Floral Parts of *Curcuma longa* L. *Asian Journal of Research in Botany* 2(2): 1-6.
40) Masunda, A. T.; Inkoto, C. L.; Bongo, G. N.; Oleko Wa Oloko, J. D.; Ngbolua, K. N.; Tshibangu, D. S. T.; Tshilanda, D. D.; Mpiana, P. T. 2019. Ethnobotanical and Ecological Studies of Plants Used in the Treatment of Diabetes in Kwango, Kongo Central and Kinshasa in the Democratic Republic of the Congo. *International Journal of Diabetes and Endocrinology* 4(1): 18-25 doi: 10.11648/j.ijde. 20190401.14.
41) Ngbolua, K. N.; Ngambika, G. K.; Amisi, C. M.; Masengo, C. A.; Gbolo, B. Z.; Ngiala Bongo, G. N., Djolu, R. D.; Mukiza, J.; Mpiana, P.T. 2019. Evaluation of Anti-malarial Drugs Use in Gbadolite Health Area in Democratic Republic of the Congo: A Cross-Sectional Study. *Bioscience and Bioengineering* 5(1): 1-10.

42) Bomanda, B. T.; Waudo, W., Ngoy,a B. P.; Muya, J. T.; Mpiana, P. T.; Mbala, M.; Openda, I.; Mack, J. and Nyokong, T. 2019. Photophysical and in vitro Antibacterial Studies of 2,6-DibromoBODIPY Dye Substituted with Dithienylenevinylene at 3,5-Positions. *Dipyrromethenes* doi: 10.6060/mhc180898n.

43) Mpiana P. T. 2019. Etude des instruments juridiques de la RD Congo sur la gestion des produits chimiques, *International Journal of Innovation and Scientific Research* 42(2): 94-103.

44) Oleko WA Oleko, J-D.; Mpiana, P. T. 2019. Oppositions to vaccination activities against polio in the province of TSHOPO / R.D. Congo; *International Journal of Medical Science Clinical Research* 1(1): 09-13.

45) Muya, J. T.; Mwanangombo, D. T; Tsalu, P. V.; Mpiana, P. T.; Tshibangu, D. S. T; Chung, H. 2019. *Conceptual DFT study of the chemical reactivity of four natural products with anti-sickling activity.* Spinger Nature applied Sciences 1:1457 https://doi.org/ 10.1007/s4245 2-019-1438-8.

46) Tshilanda, D. D.; Inkoto, C. L.; Mpongu, K.; Mata, S.; Mutwale, P. K.; Tshibangu, D. S. T.; Bongo, G. N.; Ngbolua, K. N.; Mpiana, P. T. 2019. Microscopic Studies, Phytochemical and Biological Screenings of *Ocimum canum. International Journal of Pharmacy and Chemistry* 5(5): 61-67 doi: 10.11648/j.ijpc.20190505.13.

47) Kapembo, M. L.; Al Salah, D. M. M.; Thevenon, F.; Laffite, A.; Bokolo, M. K.; Mulaji, C. K.; Mpiana, P. T.; Poté J. 2019. Prevalence of water-related diseases and groundwater (drinking-water) contamination in the suburban municipality of Mont Ngafula, Kinshasa (Democratic Republic of the Congo), *Journal of Environmental Science and Health*, Part A, doi: 10.1080/109 34529.2019.1596702.

48) Oleko WA Oleko, J. D.; Tongo R. N.; Lohohola, P. O; Likoko, J. K.; Chantal N Zingabako, C. N.; Mpiana, P. T., 2019. What epidemiological profile of breast cancer patients treated at university clinics in Kisangani, DR Congo?. *International Journal of Medical and Health Research* 5(8): 35-39.

49) Oleko Wa Oleko, J. D., Tshibanda, J. M.; Tongo, N. R. and Mpiana, P.T. 2019. Evaluation of nutritional status of children in nursery schools in Kisangani city, dr Congo. *European Journal of Pharmaceutical and Medical Research* 6(9) 59-62.
50) Gbolo, Z. B.; Ngbolua, K. N.; Mpiana, P. T.; Ndanga, B. A.; Pangodi, A. J-M.; Masengo, A. C.; Mudogo, V. 2019. Evaluation of the Clinical Efficiency of an Antisickling Polyherbal Formula Drepanoalpha in a Sickle cell disease Patient in Gbado-Lite City (Democratic Republic of the Congo) by Quantum Magnetic Resonance Analyzer. *Britain International of Exact Sciences (BIoEx) Journal,* 1 (1): 36-48.
51) Kitadi, J. M.; Lengbiye, E. M.; Gbolo, B. Z.; Inkoto, C. L.; Muanyishay, C. L.; Lufuluabo, G. L.; Tshibangu, D. S. T; Tshilanda, D. D.; Mbala, B. M.; Ngbolua, K. N.; Mpiana, P. T. 2019. *Justicia secunda* Vahl species: Phytochemistry, Pharmacology and Future Directions: A mini-review. Discovery Phytomedicine 6 (4): 157-171. doi:10.15562/phytomedicine.2019.93.
52) Lengbiye, M. E., Ngbolua, K. N.; Messi, L. M.; Mbembo wa Mbembo, B.; Bongo, G. N., Mutwale, K. P., Ngombe, K. N., Ngo Mbing, J.; Pegnyemb, D. E.; Mpiana, P. T. 2019. *In vitro* Evaluation of the Anti-scavenging and Anthelmintic Activities of *Artocarpus heterophyllus* LAM Leaves (Moraceae) in the Democratic Republic of Congo. *International Journal of Biomedical Engineering and Clinical Science* 2019; 5(2): 14-22 Doi: 10.11648/j.ijbecs.20190502.11.
53) Ngbolua, K. N.; Ngiala, G. N.; Inkoto, C. L.; Masengo, C. A.; Lufuluabo, G. L.; Janvier Mukiza, J. Mpiana, P.T., 2019. A mini-review on the Phytochemistry and Pharmacology of the medicinal plant species *Persea americana* Mill. (Lauraceae) *Discovery Phytomedicine,* 6(3): 102-111. doi:10.15562/phytomedicine. 2019.99.
54) Ngbolua, K. N.; Ngambika, G. K.; Amisi, C. M.; Ashande, C. M.; Gbolo, B.Z.; Bongo, G.N.; Djolu, R.D.; Mukiza, J.; Mpiana, P.T. 2019. Evaluation of Anti-malarial Drugs Use in Gbadolite Health Area in Democratic Republic of the Congo: A Cross-Sectional Study. *Bioscience and Bioengineering* 5 (1): 1-10.

55) Ngunde-te-Ngunde, S.; Lengbiye, E. L.; Tshidibi, D. J.; Kengo, V. F.; Djolu, R. D.; Masengo, C. A.; Gbolo, B. Z.; Iteku, B. J.; Mpiana, P. T.; Ngbolua, K.N. 2019. Antisickling and Antibacterial Activities of *Anthocleista schweinfurthii Gilg.* (Gentianaceae) from Non-human Primates Pharmacopoeia in Democratic Republic of the Congo. Budapest International Research in Exact Sciences (BirEx) Journal, 1(3): 14-20 doi: *https://doi.org/10.33258/birex. v1i3.345.*
56) Mbadiko, C. M.; Bongo, G. N.; Mindele, L. U.; Ngbolua, K. N.; Mpiana, P. T.; Ngombe, N. K.; Mutwale, P. K.; Mbemba, T. F. 2019. Effect of Drying on the Composition of Secondary Metabolites in Extracts from Floral Parts of *Curcuma longa* L. *Asian Journal of Research in Botany* 2(2): 1-6.
57) Masunda, A. T.; Inkoto, C. L.; Bongo, G. N.; Oleko Wa Oloko, J. D.; Ngbolua, K. N.; Tshibangu, D. S. T.; Tshilanda, D. D.; Mpiana, P. T. 2019. Ethnobotanical and Ecological Studies of Plants Used in the Treatment of Diabetes in Kwango, Kongo Central and Kinshasa in the Democratic Republic of the Congo. *International Journal of Diabetes and Endocrinology* 4(1): 18-25 doi: 10.11648/j.ijde. 20190401.14.
58) Ngbolua, K. N.; Ngambika, G. K.; Amisi, C. M.; Masengo, C. A.; Gbolo, B. Z.; Ngiala Bongo, G. N., Djolu, R. D.; Mukiza, J.; Mpiana, P.T. 2019. Evaluation of Anti-malarial Drugs Use in Gbadolite Health Area in Democratic Republic of the Congo: A Cross-Sectional Study. *Bioscience and Bioengineering* 5(1): 1-10.
59) Bomanda, B. T.; Waudo, W., Ngoy,a B. P.; Muya, J. T.; Mpiana, P. T.; Mbala, M.; Openda, I.; Mack, J. and Nyokong, T. 2019. Photophysical and in vitro Antibacterial Studies of 2,6-DibromoBODIPY Dye Substituted with Dithienylenevinylene at 3,5-Positions. *Dipyrromethenes* doi: 10.6060/mhc180898n.
60) Ngbolua, K. N.; Gbolo, B. Z.; Tshidibi, J. T.; Tshibangu, D. S. T.; Memvanga, P. B.; Mpiana, P. T. 2018. Effect of Storage on the Bioactivity of Drepanoalpha® (An Anti-Sickle Cell Disease Polyherbal Formula) and Comparative Biochemical Profile of Different Batches. *International Journal of Chemical and Biomolecular Science* 4(4): 60-68.

61) Ngbolua, K. N.; Ndanga, A.; Gbatea, A.; Djolu, R.; Ndaba, M.; Masengo, C.; Likolo, J.; Falanga, C.; Yangba, S.; Gbolo, B. Z., Mpiana, P. T. Environmental Impact of Wood-Energy Consumption by Households in Democratic Republic of the Congo: A Case Study of Gbadolite City, Nord-Ubangi *International Journal of Energy and Sustainable Development* 3(4): 64-71.
62) Suami, R. B.; Sivalingam, P.; Kabala, C. D.; Otamonga, J.-P.; Mulaji, C. K.; Mpiana, P. T.; Poté, J. 2018. Concentration of heavy metals in edible fishes from Atlantic Coast of Muanda, Democratic Republic of the Congo. *Journal of Food Composition and Analysis* 73:1–9.
63) Bongo, G.; Ngbolua, K. N., Ashande, C.; Gbolo, B.; Tshiama, C.; Tshilanda, D.; Tshibangu, D.; Ngombe, N.; Mbemba,T.; Mpiana, P., 2018. Antidiabetic, Antisickling and Antibacterial Activities of *Anacardium occidentale* L. (Anacardiaceae) and *Zanthoxylum rubescens* Planch. Ex Hook (Rutaceae) from DRC. *International Journal of Diabetes and Endocrinology*; 3(1): 7-14.
64) Muanyishay, C. L.; Mutwale, P. K.; Diamuini, A. N; Luhahi, F. L.; Ngombe, N. K.; Luyindula, S. N.; Mpiana, P. T., 2018. Microscopic Features, Chromatographic Fingerprints and Antioxidant Property of *Tetracera rosiflora* Gilg; *Sch. Bull.*, 4(5): 402-407.
65) Muanyishay, L. C.; Diamuini, N. A.; Mutwale, K. P.; Ngombe, K. N.; Bulubulu, F.; Luhahi, L. F.; Bongo, G.N.; Luyindula, N. S.; Mpiana, P. T., 2018. Callogenesis Induction on different types of explants of *Tetracera rosiflora* Gilg. *J. of Advancement in Medical and Life Sciences.* V7I1.02. doi: 10.5281/zenodo.140 7786.
66) Ngbolua, K. N.; Bongo, G.; Nsimba, B.; Lengbiye, E.; Iteku, J.; Kilunga, K.; Gafuene, G.; Masengo, C.A.; Tshiama, C., Kongo, N.; Inkoto, C.; Mulaji, C.; Ngombe, N.; Mbemba, T.; Mpiana, P.T. 2018. Isolation of Antibiotic Resistant Bacteria from Makelele River (Kinshasa, DR Congo) and Their Susceptibility Towards Plant-Derived Silver Nanoparticles. *International Journal of Life Science and Engineering*, 3(2) 25-38.
67) Ngbolua, K. N.; Doikasiye, J. A. E, Masengo, C. A.; Ngiala, G. B., Ngutulu, N. T., Djoza, R. D., Amédée Kundana, G. A., Mawi, C. F.;

Ngelinkoto, P. P. B.; Mpiana, P. T., 2018. Impact of Ichthyotoxic Plants on Biodiversity in the Freshwater of Businga Territory, Nord Ubangi Province in the Democratic Republic of the Congo. *International Journal of Animal Biology*, 4(4): 45-51.

68) Ngbolua, K. N.; Inkoto, C. L.; Bongo, G. N.; Lufuluabo, G. L.; Nsimba, N. K.; Masengo, C. A.; Mutanda, S. K.; Gbolo, B. Z.; Tshilanda, D. D.; Mpiana, P. T., 2018. Microscopy features, Phytochemistry and Bioactivity of *Mondia whitei* L. (Hook F.) (Apocynaceae): A mini-review. *Discovery Phytomedicine*, 5(3): 34-42.

69) Oleko Wa Oleko, J.-D.; Batina, S. A.; Hakonyange, A. O.; Lohohola, P. O.; Mwamba, H.; Tshilanda, D. and Mpiana, P. T. 2018. What Sociological Perspective of Drepanocytosis in Parents living in Kisangani, D. R. Congo? *J. of Advancement in Medical and Life Sciences.* V7I1.01. doi: 10.5281/zenodo.1405236.

70) Kafuti, Y. S.; Alemika, T. E.; Ojerinde, S. O.; Taba, K. M.; Mpiana, P. T.; Balogun, O.; Kindombe, N.M. 2018. Phytochemical studies, in vitro antioxidant and antiproliferative of the stem bark of *Boswellia dalzielii* hutch. *GPH-Journal of Advance Research in Applied Science* 1(1): 40-49.

71) Bongo, G. N; Tuntufye, H. N.; Malakalinga, J.; Ngbolua, K. N.; Pambu, A. L.; Tshiama, C.; Mbadiko, C. M.; Makengo, G. K.; Mpiana, P. T.; Mbemba, T. F.; Kazwala, R. 2018. Anti-Mycobacterial Activity on Middlebrook 7H10 Agar of Selected Congolese Medicinal Plants. *Bioscience and Bioengineering*, 4(4) 68-77.

72) Mbaya, I. E.; Beya, D. J.-P., Mansiantima, L.D.; Dzama, L.Y.; Mutombo, M.D., Beta, M.M.; Gbabete, K.J.-R., Gbolo, B.Z.; Mpiana, P.T.; Ngbolua, K.N., 2018. Fluid-Structure Interaction Modeling of Aortic Blood Flow Behavior in the Human Cardiovascular System Using Differential Equations. *International Journal of Mathematics and Computational Science*,4(3):118-123.

73) Ngbolua, K. N.; Inkoto, C. L.; Bongo, G. N.; Masengo, C. A.; Lufualabo, G. L.; Gbolo, B. Z.; Djoza, D. R.; Kwembe, J. T. K.; Onautshu, O.; Mpiana, P. T. 2018. An Updated review on the Bioactivities and Phytochemistry of the Nutraceutical Plant *Moringa*

oleifera Lam (Moringaceae) as valuable phytomedicine of multipurpose. *Discovery Phytomedicine*, 5(4): 52-63.

74) Ngbolua, K. N.; Inkoto, C. L.;Bongo, G. N.; Lufuluabo, G. L.; Nsimba, N. K.; Masengo, C. A.; Mutanda, S. K.; Gbolo, B. Z.; Tshilanda, D. D.; Mpiana, P. T. 2018. Microscopy features, Phytochemistry and Bioactivity of *Mondia whitei* L. (Hook F.) (Apocynaceae): A mini-review. *Discovery Phytomedicine* 5(3): 34-42.

75) Ngbolua, K. N.; Lufuluabo, G. L.; Lengbiye, E. M.; Bongo, G. N.; Inkoto, C. L.; Masengo, C. A.; Songowe, B. S.; Gbolo, B. Z.; Mpiana, P.T. 2018. A review on the Phytochemistry and Pharmacology of *Psidium guajava* L. (Myrtaceae) and Future direction Discovery. *Phytomedicine*,5(2): 7-13.

76) Tobotela, S. N.; Mpiana, P. T.; Nshimba, H. S. M. 2018. Ethnomycology Study of an Ectomycorhizian Mushroom Used in Cynegetic Artin Tshopo Province (Democratic Republic of the Congo) *Sch. Bull.* 4(4): 351-358.

77) Oleko wa oleko, J. D; Sembele, A.; Wetshokonda, J. J. and Mpiana, P.T., 2018. Assessment of the reasons of the cholera upsurge in Lubunga (Province of the Tshopo, Democratic Republic of Congo) *International Journal of Innovation and Scientific Research* 35(2): 58-68.

78) Oleko wa Oleko, J. D.; Tshilanda, D.; Mwamba, H.; Sabiti, M.; Batina, S.A.; Bontambo, P.N. and Mpiana, P.T. 2018. Evaluation of The Food Security of Households Living in Tshumbe Town, DR Congo. *J. of Advancement in Medical and Life Sciences.* V6I3-02. doi: 10.5281/zenodo.1195581.

79) Kabena, N. O.; Katunda, M. R.; Bikandu, K. B.; Botefa, I. C.; Ngombe, K.N., Mpiana, P.T.; Mboloko, E.J.; Lukoki, L.F. 2018. Ethnobotanical study of plants used by Pygmies for reproductive health in Mbandaka and surrounding areas / Equateur Province, DR Congo. *International Journal of Innovation and Scientific Research.* 36(1): 19-29.

80) Ngbolua, K. N.; Mpiana, P.T. and Gushimana Y. 2018. Computational Analysis of Thermal Denaturation of Human Hemoglobin by Non

Linear Regression of Gushimana Yav Equation Using Origin Software Package. *J. of Advancement in Medical and Life Sciences.* V6I3-01. doi: 10.5281/zenodo. 1195578.

81) Ngbolua, K. N.; Bongo, N. G.; Domondo, A.; Nsimba, B.; Iteku, J.; Lengbiye, E.; Ashande, C.; Tshiama, C.; Inkoto, C.; Lufuluabo, L.; Kilunga, K.; Gafuene, G.; Mulaji, C.; Mbemba, T.; Poté, J.; Mpiana, P.T. 2018. Synthesis and Bioactivity of Silver Nanoparticles Against Bacteria (*E. coli* and *Enterococcus* sp.) Isolated from Kalamu River, Kinshasa City, *Democratic Republic of the Congo Frontiers in Environmental Microbiology* 4(1): 29-40.

82) Lengbiye, E. M.; Ngbolua, K. N.; Bongo, N. G.; Messi, L. M.; Noté, O.P.; Mbing, J.N.; Pegnyemb, D.E. and Mpiana, P.T. 2018. *Vitex madiensis Oliv.* (Lamiaceae): phytochemistry, pharmacology and future directions, a mini-review. *Journal of Pharmacognosy and Phytochemistry* 7(2): 244-251.

83) Mutimana, K R ; Muya, J. T.; Mudogo, V.; Muzomwe, M.; Mpiana, P.T. 2018. Theoretical study of the regioselectivity of the interaction of molecules isolated from *Siphonochilus aethiopicus* with water, *Scholars Bulletin* 4(2): 222-230.

84) Mbula, J. P.; Kwembe, J. T. K.; Tshilanda, D. D.; Ngobua, K. N.; Mbala, B. M.; Nsimba, S. M.; Onautshu, O.; Mpiana, P. T. 2018. Antihemolytic, radical scavenging and antibacterial activities of essential oil of *Fagara macrophylla* (Oliv) Engl from Masako forest reserve (RD Congo) *Sch. Bull.*, 4 (1): 74-80.

85) Mbula, J. P.; Kwembe, J. T. K.; Tshilanda, D. D.; Ngobua, K. N.; Kabena, O. N.; Nsimba, S. M.; Onautshu, O.; Mpiana, P. T. 2018. Antisickling, antihemolytic and radical scavenging activities of essential oil from *Entandrophragma Cylindricum* (Sprague) Sprague (Meliaceae). *J. of Advancement in Medical and Life Sciences.* V6I2-03. doi: 10.5281/zenodo.1167931.

86) Ngbolua, K. N.; Tshibangu, D. S. T.; Mpiana, T.P.; Mudogo, V.; Tshilanda, D. D.; Ashande, C. M.; Divakar, S.D.; Ramanathan, M.; Syamala, G. 2018. Medicinal Plants from Democratic Republic of the

Congo as Sources of Anticancer Drugs. *J. Prev. Med:* JPVM-102. DOI: 10.29011/JPVM-102. 100002.

87) Ngbolua, K. N.; Tshibangu, D.S.T.; Mpiana, P.T.; Mudogo, V.; Tshilanda, D.D.; Ashande, C.M.; Divakar, S.; Ramanathan, M.; Syamala, G. 2018. Medicinal Plants from Democratic Republic of the Congo as Sources of Anticancer Drugs. *J. of Advanced Botany and Zoology.* V6I1.01. DOI: 10.5281/zenodo.1162973.

88) Inkoto, C. L.; Bongo, G. N.; Mutwale, K. P.; Masengo, A. C.; Gbolo, B. Z.; Tshiama, C.; Ngombe, N. K.; Iteku, B. J.; Mbemba, T. M. Mpiana, P.T.; Ngbolua, K.N. 2018. Microscopic features and chromatographic fingerprints of selected congolese medicinal plants: *Aframomum alboviolaceum* (Ridley) K. Schum, *Annona senegalensis* Pers. and *Mondia whitei* (Hook.f.) *Skeels Emer Life Sci. Res.* 4(1): 1-10 doi: http://dx.doi.org/10.7324/ELSR. 2018.410110.

In: *Ocimum basilicum*
Editor: Andres A. Walton

ISBN: 978-1-53619-265-0
© 2021 Nova Science Publishers, Inc.

Chapter 5

ANTISICKLING ACTIVITY OF *OCIMUM BASILICUM* AND SOME OF ITS COMPOUNDS

*Dorothée D. Tshilanda[1], Carlos N. Kubengele[1],
Etienne M. Ngoyi[1], Aristote Matondo[1],
Jason T. Kilembe[1], Giresse N. Kasiama[1],
Clement L. Inkoto[2], Emmanuel M. Legbiye[2],
Benjamin Z. Gbolo[2,3], Gédéon N. Bongo[2],
Damien S. T. Tshibangu[1], Koto-te-Nyiwa Ngbolua[2,3]
and Pius T. Mpiana[1],**

[1]Department of Chemistry, Faculty of Sciences,
University of Kinshasa, Kinshasa XI,
Democratic Republic of the Congo
[2]Department of Biology, Faculty of Sciences,
University of Kinshasa, Kinshasa XI,
Democratic Republic of the Congo
[3]Department of Basic Sciences, Faculty of Medicine,
University of Gbado-Lite, Gbado-Lite,
Democratic Republic of the Congo

* Corresponding Author's E-mail: ptmpiana@gmail.com.

Abstract

Sickle cell disease is an inherited disorder characterized by a structural abnormality of hemoglobin S. According to the WHO, several thousand people worldwide suffer from this disease, with the majority of cases recorded in Africa. The ineffectiveness and inaccessibility of the treatments presented, force populations to resort to medicinal plants. Thus, in addition to its culinary use as a spice, *Ocimum basilicum* is also cited in the treatment of various diseases such as cancer, diarrhea, hemorrhoids, rhinitis including sickle cell disease.

The antisickling activity of the acidified methanol extract, anthocyanin fraction and essential oil as well as that of butyl stearate and rosmarinic acid two pure compounds isolated from Ocimum basilicum L was evaluated using the Emmel test. The results showed that all these extracts and pure compound exhibited an interesting antisickling activity. These activities are dose-dependent with normalization rate higher than 80%. Rosmarinic acid and methanolic extract showed a minimum normalization concentrations of 0.18 ± 0.03 mg/mL and 0.23 ± 0.04 mg/mL respectively. Essential oil showed good antioxidant activity with $IC_{50} = 1.180 + 0.015$ using the DPPH test.

Keywords: *Ocimum basilicum*, antisickling activity, Rosmarinic acid, butyl stearate, antioxidant activity

1. Introduction

Sickle cell disease is characterized by a change in the shape of the red blood cells from the biconcave form to a sickle form. Sickle-cells can clog small blood vessels and block blood flow, leading to acute chronic pain, severe bacterial infections and necrosis (Lewis 2015; Mpiana et al. 2016).

According to the World Health Organization (WHO), approximately 330,000 infants are born with hemoglobin disorders each year and nearly five million people worldwide are affected by this disease with many of them in Africa. In the Democratic Republic of Congo (DRC), about two million people are affected by sickle cell disease, or two percent (homozygous SS) of the population. As the percentage of people affected continues to increase, sickle cell disease has become a real public health problem in the homes of people living in developing countries(WHO, 2020; Muya et al. 2019).

Several treatment options have been developed to alleviate patients, including bone marrow transplantation, gene therapy, repeated blood transfusions, hydroxyurea etc. (Lewis 2015). However, it has been shown that these treatments are not only ineffective in some cases, but also very costly for poor populations in Africa. In addition, blood transfusion exposes patients to the risk of HIV/AIDS infections in this continent (Kitadi et al. 2020).

To face these limitations, the use of medicinal plants with antisickling activity remains the only means readily available to these populations. Indeed, plants have the capacity to produce a large number of secondary metabolites with interesting therapeutic properties (Mpiana et al. 2020; Ngbolua et al. 2020).

O. basilicum, an aromatic plant of the Lamiacea family is widely used in traditional medicine for its anti-inflammatory, anti-hyperglycemic, antioxidant and antisickling properties (Thiraviyam et al. 2019; Tshilanda et al. 2014). It contains many vitamins, mineral salts and is used in DRC as a spice and in different traditional recipes against hemorrhoids, diarrhea, women's intimate hygiene (Pachkore and Dhale 2012; Tshilanda et al. 2016a,b; Kabena et al. 2014).

Thus, the objectives of the present work is to study the antisickling and antioxidant activities of extracts, essential oils and different pure molecules isolated or identified from *O. basilicum*.

2. MATERIAL AND METHODS

2.1. Material

2.1.1. Plant Material

The leaves of *O. basilicum* were collected in the garden around the University of Kinshasa in the commune of Lemba, in the city of Kinshasa (DRC). These samples were identified at the herbarium of the National

Institute of Agronomic Studies and Researches (INERA), at the Faculty of Sciences of the University of Kinshasa. They were dried at room temperature in the shade for two weeks, then powdered with an electric grinder. The essential oil has been extracted from the fresh material.

2.1.2. Blood Samples

The blood samples analyzed were taken from several sickle cell patients at the «Centre de Médecine Mixte et d'Anémie SS» located in Kinshasa area, DRC and stored at 4°C in a refrigerator.

2.2. Methods

2.2.1. Phytochemical Studies

2.2.1.1. Phytochemical Screening
a) Screening in solution

Chemical screening was performed in aqueous and organic extracts according to well-known protocols (Bruneton 1999).

b) Screening by TLC and HPLC

Analytical thin layer chromatography (TLC) was carried out on 10µL of the methanolic extract with the elution system: dichloromethane/ acetone/formic acid/(85: 25: 8.5). Rosmarinic acid and caffeic acid were used as standards. After development, the plaque was visualized at 366nm after using the natural developer PEG-reagent. This TLC was completed by high performance liquid chromatography (HPLC). The HPLC analysis was carried out at 25°C by an "Agilent 1100" HPLC chain connected to a diode array detector (DAD) as described in our previous work (Tshilanda et al. 2016b).

2.2.1.2. Anthocyanin Extract Preparation

The powder of *O. basilicum* (300g) was placed in a flask containing 1L of methanol acidified with 0.4mol/L of hydrochloric acid, then concentrated at reduced pressure using a rotavapor. The concentrate was delipidated using petroleum ether and then dried at 50°C in an oven (Brand MEMMERT model).

2.2.1.3. Fractionation of the Acidified Methanolic Extract

The thin layer chromatography with silica gel was carried out on the acidified methanolic extract using the mixture of n-butanol-acetic acid-water (4-1-5) as eluting system and the UV lamp (type CAMAG) at 366 nm as a developer (Tshilanda et al. 2014, Tshilanda et al. 2016a,b).

Preparative chromatography was performed on glass plates 20cm x 20cm on which was spread P/UV254 silica gel. These plates were dried at 105°C for 48h in the oven.

Column chromatography was then carried out to isolate the mixtures from the preparative chromatography, using silica gel (Kieselgel brand 60 F254; 0.2 - 0.5 nm/35 - 70 100 mesh) and the mixture of n-butanol/n-hexane (8:2) as eluting system.

2.2.1.4. Spectroscopic Analyses

The structure elucidation of the compound isolated from the extract was done using 1D-NMR (^{1}H-NMR, ^{13}C-NMR), 2DNMR (COSY, HMBC) and mass spectrum at high resolution. NMR spectra were recorded using the spectrometer Bruker Avance 300 MHz type. All spectra were taken at room temperature.

2.2.1.5. Essential Oil Extraction and Analysis

Essential oil has been produced by hydro-distillation as earlier reported (Tshilanda et al. 2016a).

A weighed amount of leaves was immersed in a 500 mL round bottom flask of water and hydrodistilled. Water and essence were recovered in a decant bowl, and anhydrous magnesium sulfate was used for drying trace of water.

Oil was stored in a dark glass bottle at 4°C before gas chromatography (GC) and gas chromatography-mass spectrometer (GC-MS) analyses as previously described (Tshilanda et al. 2016a).

2.2.2. Biological Activities

2.2.2.1. Antisickling Test

Antisickling bioassy was done using Emmel test as described in our previous work (Mpiana et al. 2016, Mpiana et al. 2008, Bongo et al. 2017). Briefly, the blood sample was mixed with plant extracts at different concentrations using physiological saline (0.9% NaCl) as dissolution solvent. The control consists of diluted sickle cell blood wihout extract. The effect of the various extracts is observed by optical microscopy after an exposure time of 24 and 48 hours under hypoxia and isotonic conditions. A digital camera was used to record microscopic images of erythrocytes obtained. These micrographs were then processed by computer software MOTIC pictures 2000 version 1.3 and dose-type curves obtained with the Origin 8.5 Pro software.

2.2.2.2. DPPH Radical Scavenging Activity

The 2,2-diphényl-1-picrylhydrazyle (DPPH) free radical scavenging assay was carried out as previously reported (Mpiana et al. 2015). About 3.5 mL of 0.3 mmol/L solution of DPPH radical in methanol were added to 0.5 mL solution of essential oil. Bioactive essential oil solutions were used at the same values of concentration in methanol for comparison. Each mixture was submitted to spectrophotometry (UV-vis 320/Safas Monaco Spectrophotometer) analysis.

Mixture of essential oil solution with DPPH radical solution in methanol were shaken vigorously and absorbances were recorded at 517nm during 35min as equilibrium time. DPPH radical scavenging has been determined by percentage of reduction that provides IC_{50} (concentration of essential oil or ascorbic acid that reduce 50% of DPPH radical concentration) by extrapolation as antioxidant effectiveness (Ngbolua et al. 2014). Ascorbic acid was tested as standard for comparisons. Percentages of reduction were calculated according to the following equation:

$$\% = [1 - \frac{A_{sample}}{A_{blank}}] \times 100$$

where A_{blank} is absorbance of blank and A_{sample} is absorbance of the tested extract.

The IC_{50} is determined by plotting the percentage reduction of DPPH as a function of the content of the antioxidant substance.

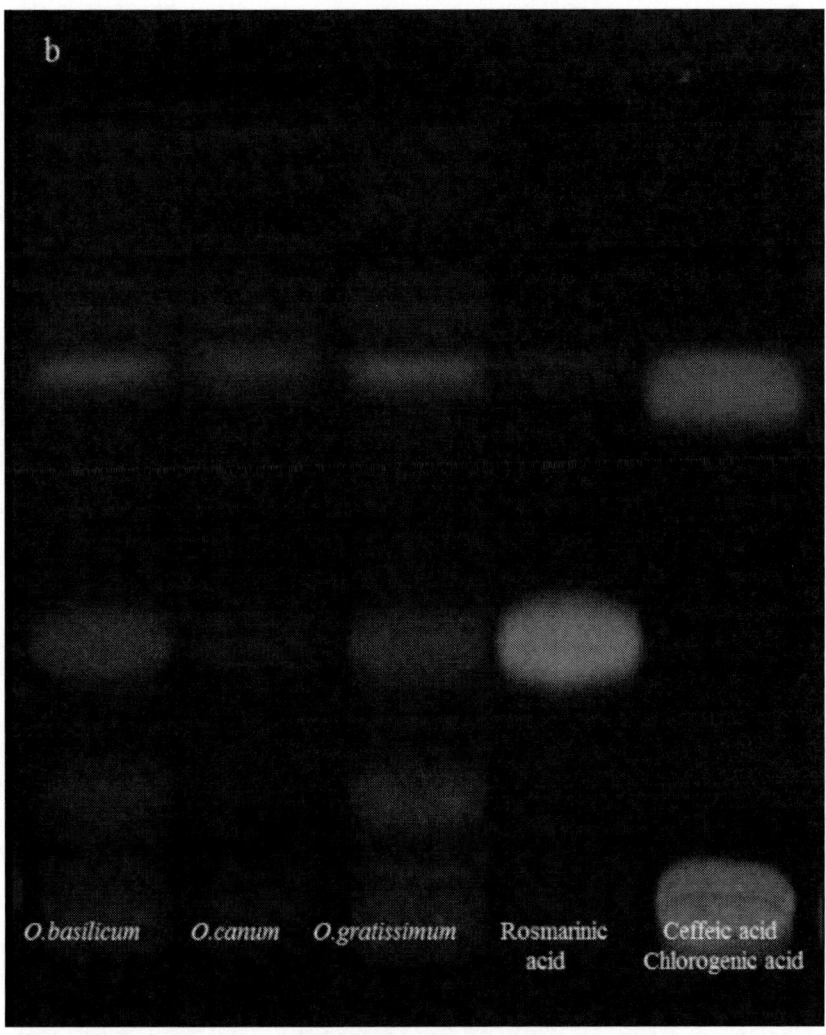

Figure 1. TLC Chromatogram of methanolic extracts from *O. basilicum* with rosmarinic acid and caffeic acid as standards; developed with Methylene Dichloride/acetone/formic acid 85:25:8.5 and visualized at 365nm with Natural Products-PEG reagent. Two other species of Ocimum (*O. canum et O. gratissimum*) are used for comparison.

3. Results and Discussion

3.1. Thin Layer Chromatography Analysis

Figure 1 gives the chromatogram of the methanolic extract of *O. basilicum* at 366nm in the presence of rosmarinic and caffeic acids as standard. The extract was, for the sake of comparison, next to that of two other *Ocimum* species (*O. gratissimum* and *O. canum*).

This figure shows that *O. basilicum* contains rosmarinic acid as well as *O. gratissimum* and *O. canum*. This was confirmed by HPLC analysis

3.2. High Performance Liquid Chromatography Analysis

The phytochemical analysis carried out by High Performance Liquid Chromatography (HPLC) on the methanolic extract of *O. basilicum* revealed rosmarinic acid as a major component thus confirming the results of TLC and our previous work (Tshilanda et al. 2016b).

In fact, the HPLC fingerprint of pure rosmarinic used as control for comparison to the methanolic extracts of *O. basilicum* HPLC fingerprints showed that the pure rosmarinic acid appears as a single peak to the retention time of 32.408 minutes. In the fingerprint of *O. basilicum* the most intense peak at almost the same retention time (32.300min) as already found in other previous works with *O. basilicum* samples from other countries (Vlase et al. 2014).

3.3. Spectroscopic Analysis of Methanolic Extract

Chromatographic analysis on the methanolic extract allowed to obtain at Rf: 0.71, a viscous brown product, fluorescent at UV lamp at 366nm, with a b.p. of 222 - 224°C. ^{1}H-NMR spectrum (CDCl$_3$) showed characteristic peaks at δ_H: 4.064ppm (t), 2.28ppm (t), 0.93ppm (t), 0.88ppm

(t) and 1.65ppm to 1.31ppm. The ^{13}C-NMR spectrum showed 15 characteristic peaks at δ_C :174.0ppm, 64.1ppm, 34.4ppm, 31.9ppm, 30.7ppm, 29.7ppm, 29.6ppm, 29.5ppm, 29.3ppm, 29.2ppm, 25.0ppm, 22.1ppm, 19.1ppm, 14.1ppm, 13.7ppm and a characteristic peak of great intensity at 29.7ppm.

The mass spectrum, performed in electron impact ionization positive mode showed the molecular ion peak M$^{+\cdot}$ at m/z: 340.3345 and other peaks respectively at: 312.3, 257.2, 239.2, 213.18 and 199.17. These spectral characteristics, according to the literature correspond to the butyl stearate (Tshilanda et al. 2014).

This compound was synthesized and presented the same spectroscopic characteristics than the natural one.

3.4. Gas Chromatography and GC-MS Analysis of Essentiel Oils

A total of 40 compounds (99.98%) were identified by GC, and only 35 compounds (98.10%) were characterized by coupled GC–MS (MS and RI), and five compounds (1.88%) were unidentified. Chemical composition of this essential oil is mainly dominated by terpenes and hydrocarbon compounds. The overall composition (99.98%) can be described as follow: non-oxygenated monoterpenes (0.67%), oxygenated monoterpenes (26.75%), nonoxygenated sesquiterpenes (14.41%), oxygenated sesquiterpenes (10.21%), oxygenated hydrocarbons (0.04%), aromatic hydrocarbons (0.02%), and oxygenated aromatic hydrocarbons (46.00%). The remaining compounds (1.88%) were unidentified among spectra stored in Wiley Registry 10th libraries. Major compounds (> 1.50 %) of this chemical composition are methyl chavicol or estragole (35.72%), linalool (21.25%), epia- cadinol (8.02%), a-bergamotene (6.56%), eugenol (4.60%), 1,8- cineole (4.04%), germacrene D (2.06%), thymol (1.64%), and (E)-citral (1.55%). They represent 84.44% of the overall composition of oil; while minor compounds (0.5% – 1.5%) are (Z)-citral (1.40%), alloaromadendrene (1.30%), (E)-a-bisabolene (1.27%), d-cadinene (1.23%), nerol (1.25%), geraniol (1.01%), 1,10-di-epicubenol (0.95%), a-humulene (0.85%), b-eudesmol (0.72%), ciscaryophyllene (0.61%), a-copaene (0.53%), a-eudesmol (0.52%), camphre (0.23%), b-pinene (0.20%), trans-ocimene (0.16%), apinene (0.13%), camphene (0.04%), oct-1-en-3-ol (0.04%), sabinene (0.03%), myrcene (0.03%), a-thujene (0.02%), p-cymene (0.02%), g-terpinene (0.02%), cis-sabinene hydrated (0.02%), dterpineol (0.02%), terpineol-4 (0.02%).

This chemical composition reveals that essential oil of *O. basilicum* growing in DRC is rich in oxygenated aromatic hydrocarbons (46.00%) and oxygenated monoterpenes (26.75%). The occurrence of methyl chavicol (35.72%) and linalool (21.25%) as most abundant compounds indicates that the chemo-type of this essential oil is linalool-methyl chavicol.

The oxygenated and non-oxygenated sesquiterpenes represent 14.41% and 10.21%, respectively. Only 0.67% of nonoxygenated monoterpene was identified. The unidentified compounds and aromatic hydrocarbons represent 1.88% and 0.04%, respectively. This confirms our previous work (Tshilanda et al. 2016a).

3.5. Biological Activities

3.5.1. Antisickling Activities

Figures 2 to 7 provide digital images of respectively SS blood alone (negative control), of the SS blood treated with acidified methanol extract, butyl stearate, rosmarinic acid, anthocyanin fraction and essential oil.

Figure 2 shows that for the untreated blood (negative control), the majority red blood cells are sickleshaped, confirming the Sickle cell nature of the blood.

However, when the Sickle red blood cells are treated with acidified methanol extract, butyl stearate, rosmarinic acid, anthocyanin fraction, and essentail oil, the majority of the erythrocytes recovered a normal shape. This indicates that these extracts and molecules have antisickling activity and show also a good normalization effect of drepanocytes in hypoxic condition.

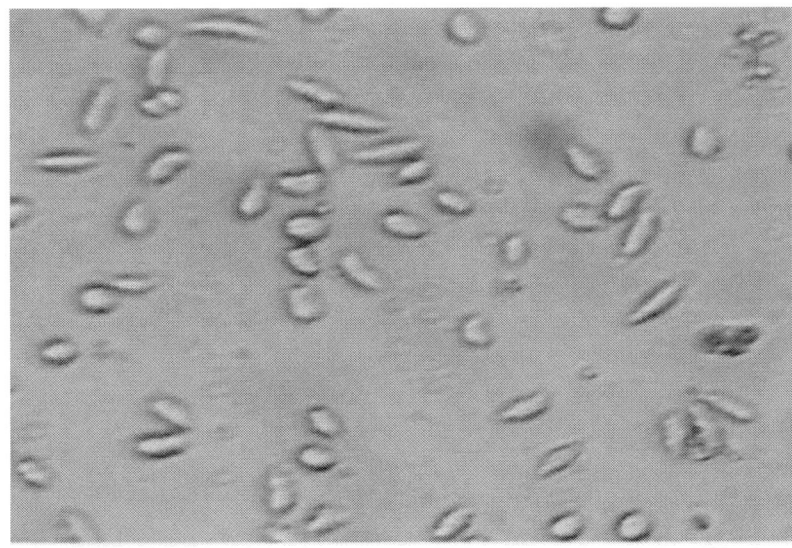

Figure 2. Morphology of drepanocytes of SS blood (Negative control) [NaCl 0.9%, $Na_2S_2O_5$ 2%] (500X).

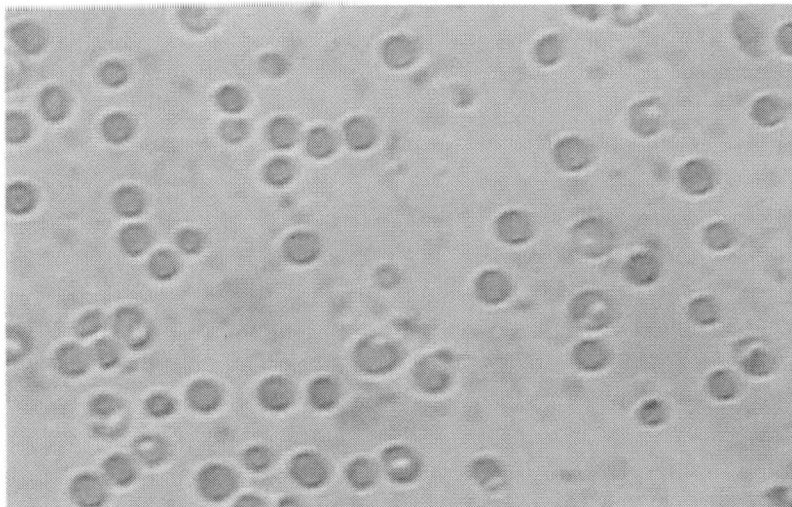

Figure 3. Morphology of drepanocytes treated with acidified methanolic extract [NaCl 0.9%, $Na_2S_2O_5$ 2%] (500X).

The same behavior was already observed for active extracts from some plants used in the management of sickle cell disease in Congolese traditional medicine and their extracts justifying their traditional use.

These molecules and others contained in these extracts would interact with hemoglobin S (HbS) and compete with its polymerization reaction.

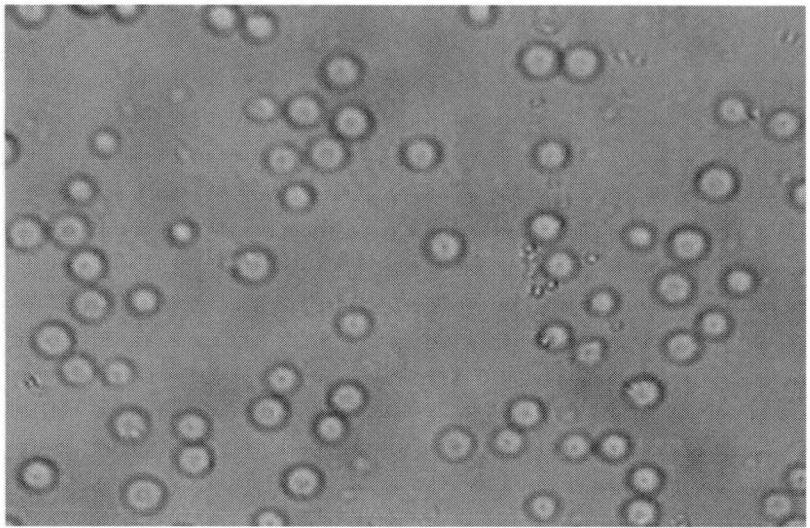

Figure 4. Morphology of drepanocytes treated with butyl stearate [NaCl 0.9%, $Na_2S_2O_5$ 2%] (500X).

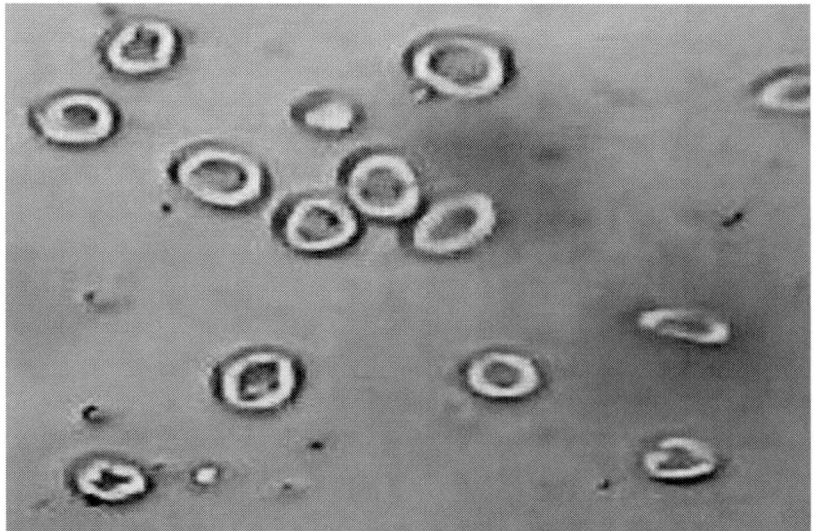

Figure 5. Morphology of drepanocytes treated with rosmarinic acid [NaCl 0.9%, $Na_2S_2O_5$ 2%] (500X).

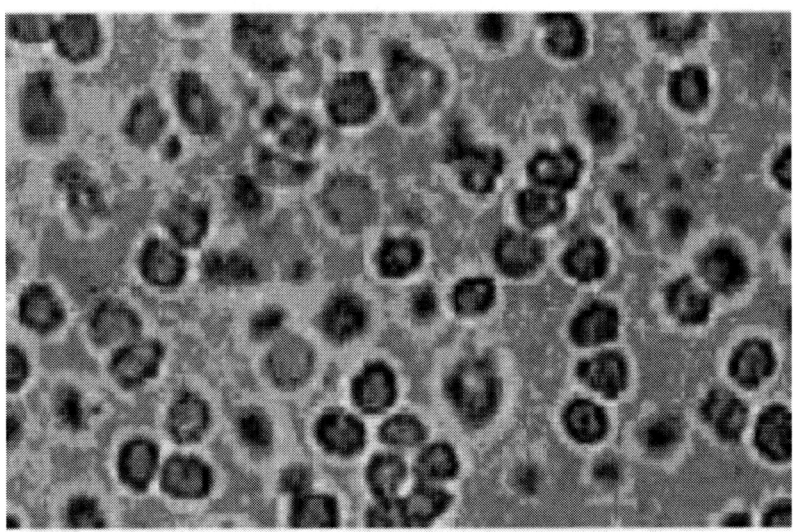

Figure 6. Morphology of drepanocytes treated with anthocyanin fraction [NaCl 0.9%, $Na_2S_2O_5$ 2%] (500X).

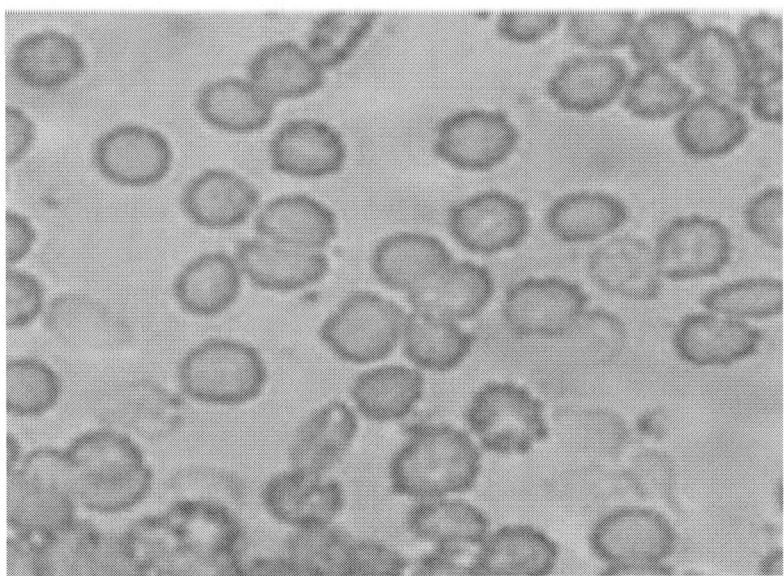

Figure 7. Morphology of drepanocytes treated with essential oil [NaCl 0.9%, $Na_2S_2O_5$ 2%] (500X).

This would prevent the sickling of red blood cells (Mpiana et al. 2007, Mpiana et al. 2008; Mpiana et al. 2015; Mpiana et al. 2016; Ngbolua et al. 2014). In fact, according to Russu et al. (1989) when investigating on molecular basis for the antisickling activity of aromatic amino acids and related compounds by NMR, some molecules like phenolic acids or their derivatives would interact with HbS on a binding site located at or near the beta 6 position (the site containing the mutation in beta 6Glu to Val). This binding would induce conformational changes in the amino-terminal domains of the beta chains.

In addition, our previous works (Shode et al. 2014; Tshilanda et al. 2016b) showed that somme terpenic acids like betulinic and maslinic acids and some esters including butyl stearate have antisickling activity and that this activity could involve both binding to deoxyhemoglobin S and modification of the erythrocyte membrane.

Quantitative parameter of antisickling activity can be given by minimal concentration of normalization (MCN) or the concentration that normalizes 50% of drepanocytes (ED_{50}). Figures 8 and 9 give as illustration, the evolution of the normalization rate of blood sickle cells form with the methanolic extracts of *O. basilicum* and rosmarinic acid (RA).

These figures show that the antisickling activity of methanolic extract and rosmarinic acid varies with concentration and is therefore dose-dependent. Buthyl stereate and anthocyanin extract also have the same behavior. This evolution of antisickling activity with concentration confirms the results already obtained in our previous studies (Mpiana et al. 2016; Tshilanda et al. 2014; Tshilanda et al. 2016b; Ngbolua et al. 2014). The values of the minimum normalization concentrations, i.e., the smallest concentration at which the product reaches its maximum rate of normalization are 0.18 ± 0.03 mg/mL and 0.23 ± 0.04 mg/mL for rosmarinic acid and ethanolic extract respectively.

Rosmarinic acid showed better antisickling activity than methanolic extract of *O. basilicum*, thus it could be the main molecule responsible for thisl activity in *O. basilicum* (Tshilanda et al. 2016b).

Antisickling Activity of Ocimum basilicum ... 127

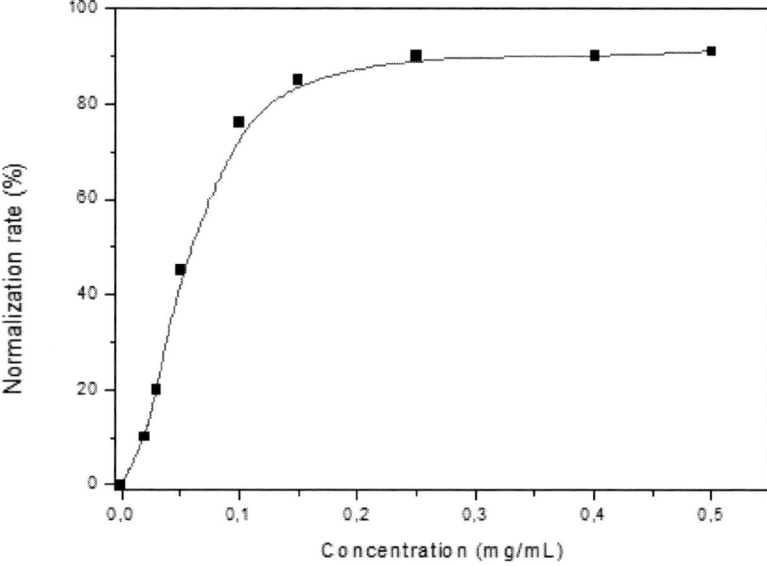

Figure 8. Evolution of the normalization rate of the drepanocytes form with the methanolic extracts concentration.

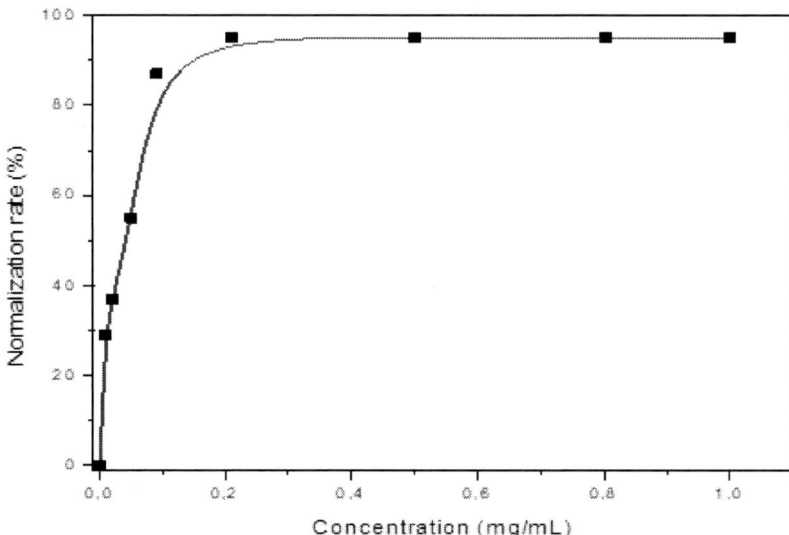

Figure 9. Evolution of the normalization rate of the drepanocytes form with rosmarinic acid concentration.

3.5.2. DPPH Radical Reduction of Essential Oil

Figure 10 gives the percentages of reduction of the DPPH radical versus concentration values of essential oil of *O. basilicum L.* using ascorbic acid as control.

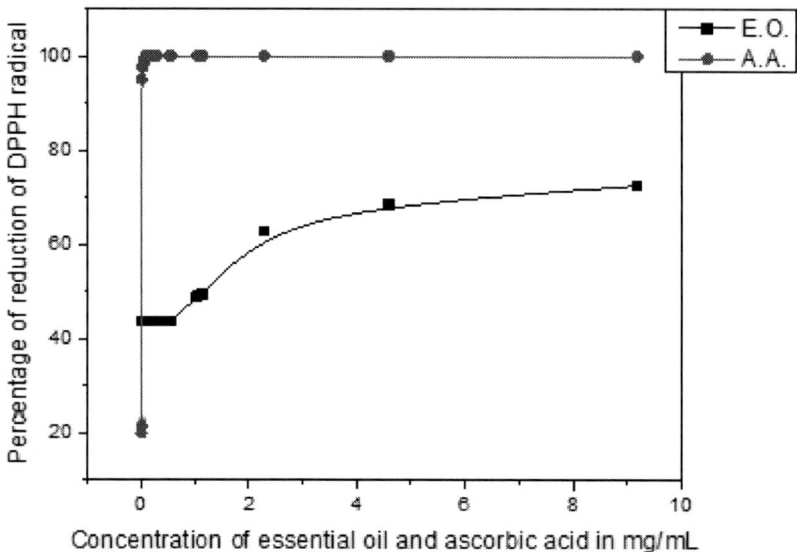

Figure 10. Percentages of reduction of the DPPH radical with concentration with *O. basilicum* essential oil (E.O.) and ascorbic acid (A.A).

As it can be seen in the figure 10, essential oil of *O. basilicum* acts as an antioxidant, in scavenging radical activity of DPPH. But its activity is slightly low to that of ascorbic acid which is used as positive control. Indeed, ascorbic acid reduces DPPH radical nearest 100% at the lowest values of concentrations and the essential oil acting at the same concentrations values reaches 80% in the used concentration. The calculated IC_{50} of Essential oil is 1.180 ± 0.015. Equivalent studies of antioxidant scavenging activity carried out by others authors in which butylated hydroxytoluene (BHT) a known antioxidant used in food has been used as control has shown that essential oil was less active than BHT (Tshilanda et al. 2014; Tshilanda et al. 2015).

It was shown that sickle cell patients are submitted to oxidative stress. Indeed, increasing evidence accumulated over the last decade indicates that reactive oxygen species (ROS) play a key role in the pathophysiology of various ischemic diseases including sickle cell disease. The oxidative stress in sickle cell patient is likely the result of intravascular sickling and transient vaso-occlusive event leading to the decrease of nitric oxide (NO) probably due to consumption of NO by free oxygen radicals, and/or by cell-free plasma heme as a result of hemolysis (Ngbolua et al. 2014). The antioxydant activity of *O. basilicum* essential oil could permit this aromatic plant to prevent *in vivo* oxidative reactions, often by scavenging ROS before they can damage cells.

CONCLUSION

The objective of this work was to evaluate the antisickling activity of *Ocimu basilicum* and some of its compounds. The results obtained show that the methanolic extract, but also the anthocyanin fraction, rosmarinic acid and butyl stereate obtained from this aromatic plant show an antisickling activity which can prevent sickle red blood cells from taking the abnormal shape responsible for the problems of sickle cell disease patients. In addition, the essential oil extracted from this plant shows an anti-free radical activity that can prevent the oxidative stress that is important in this disease and thus prevent the probable damage to the membranes of red blood cells. The combination of these biological properties would therefore be the basis of the use of this plant in traditional medicine against drepanocytosis in the DR Congo.

The determination of the structures of other compounds in this plant and the determination of their biological activity is ongoing.

REFERENCES

Bongo, G., Inkoto, C., Masengo, C., Tshiama, C., Lengbiye, E., Djolu, R., Mutwale, K., Kabamba, K., Mbemba, T., Tshilanda, D., Mpiana, P. T., Ngbolua, K. N. (2017). Assessment of Antisickling, Antioxidant and Antibacterial Activities of Some Congolese Taxa: *Aframomum alboviolaceum* (Ridley) K. Schum, *Annona senegalensis* Pers. and *Mondia Whitei* (Hook. f.) Skeels. *American Journal of Laboratory Medicine*, 2(IV): 52 - 59. Doi: 10.11648/j.ajlm.20170204.13.

Bruneton, J. (1999). *Pharmacognosie, Phytochimie et Plantes médicinales. Edition Technique et Documentation-Lavoisier, 3e édition, Paris: France.* [*Pharmacognosy, Phytochemistry and Medicinal Plants. Technical and Documentation Edition-Lavoisier*, 3rd edition, Paris: France].

Kabena, N. O., Ngombe, K. N., Ngbolua, K. N., Kikufi, B. A., Lassa, L., Mboloko, E. J., Mpiana, P. T. and Lukoki, L. F. (2014). Etudes ethnobotanique et écologique des plantes d'hygiène intime féminine utilisées à Kinshasa (République Démocratique du Congo), *Int. J. Biol. Chem. Sci.*, 8(VI): 2626 - 2642 [Ethnobotanical and ecological studies of feminine intimate hygiene plants used in Kinshasa (Democratic Republic of Congo), *Int. J. Biol. Chem. Sci.*, 8 (VI): 2626 - 2642].

Kitadi, J. M., Inkoto, C. L., Lengbiye, E. M., Tshibangu, D. S. T., Tshilanda, D. D., Ngbolua, K. N., Taba, K. M., Mbala, B. M., Schmitz, B. and Mpiana, P. T. (2020). Mineral Content and Antisickling Activity of *Annona senegalensis, Alchornea cordifolia* and *Vigna unguiculata* Used in the Management of Sickle Cell Disease in the Kwilu Province (Congo, DR) *International Blood Research and Reviews,* 11(III): 18 - 27. DOI: 10.9734/IBRR/2020/v11i330131.

Lewis, M. E.(2015). *Sickle cell disease genetics, management and prognosis.* Nova Science Publishers, Inc. New york.

Mpiana, P. T., Ngbolua, K. N., Tshibangu, D. S. T., Mwanangombo, D. T. and Tsalu, P. V. (2015). Antisickling and Radical Scavenging Activities of Anthocyanin Extracts from the Leaves of *Gardenia*

Ternifolia Subsp. Jovis-Tonantis (Welw.) Verdc. (Rubiaceae) In: *Sickle Cell Disease: Genetics, Management and Prognosis*. Nova Scienes Publishers, Inc, New York.

Mpiana, P. T., Mudogo, V., Tshibangua, D. S. T., Kitwa, E. K., Kanangila, A. B., Lumbu, J. B. S, Ngbolua, K. N., Atibu, E. K, Kakule, M. K. (2008). Antisickling activity of anthocyanins from *Bombax pentadrum*, *Ficus capensis* and *Ziziphus mucronata*: Photodegradation effect. *Journal of Ethnopharmacology*, 120: 413 - 418.

Mpiana, P. T., Ngbolua, K. N., Tshibangu, S. T. D. (2016). Les alicaments et la drepanocytose : une mini-revue. une mini-revue, *Comptes Rendus Chimie*, 19(VII):884 - 889. http://dx.doi.org/10.1016/ j.crci.2016.02.019 [Foods and sickle cell disease: a mini-review. *Comptes Rendus Chimie*, 19(VII):884 - 889].

Mpiana, P. T., Ngbolua, K. N., Tshibangu, D. S. T., Kilembe, J. T., Gbolo, B. Z., Mwanangombo, D. T., Inkoto, C. L., Lengbiye, E. M., Mbadiko, C. M., Matondo, A., Bongo, G. N., Tshilanda, D. D. (2020). Identification of potential inhibitors of SARS-CoV-2 main protease from *Aloe vera* compounds: A molecular docking study. *Chemical Physics Letters*. 754: 137751 https://doi.org/10.1016/ j.cplett.2020.137751.

Mpiana, P. T., Tshibangu, D. S. T., Shetonde, O. M., Ngbolua, K. N. (2007). In vitro antidrepanocytary activity (anti-sickle cell anemia) of some congolese plants, *Phytomedecine*, 14:192 - 195.

Muya, J. T., Mwanangombo, D. T., Tsalu, P. V., Mpiana, P. T., Tshibangu, D. S. T., Chung, H. (2019). Conceptual DFT study of the chemical reactivity of four natural products with anti-sickling activity. *Spinger Nature applied Sciences*, 1:1457. https://doi.org/10.1007/s4245 2-019-1438-8.

Ngbolua, K. N., Mpiana, P. T., Tshibangu, D. S. T., Mazasa, P. P., Gbolo, B. Z., Atibu, E. K., Kadima, J. N., Kasali, F. M. (2014). *In vitro* antisickling and radical scavenging activities of a poly-herbal formula (Drepanoalpha®) in sickle cell erythrocyte and acute toxicity study in

Wistar albino rats. *European Journal of Medicinal Plants*, 4(X):1251 - 1267.

Ngbolua, K. N., Mbadiko, C. M., Matondo, A., Bongo, G. N., Inkoto, C. L., Gbolo, B. Z., Lengbiye, E. M., Kilembe, J. T., Mwanangombo, D. T., Ngoyi, E. M., Falanga, C. M., Tshibangu, D. S. T., Tshilanda, D. D., and Mpiana, P. T. (2020). Review on Ethno-botany, Virucidal Activity,Phytochemistry and Toxicology of *Solanum* genus: Potential Bio-resources for the Therapeutic Management of Covid-19. *European Journal of Nutrition and Food Safety,* 12(VII): 35 - 48 DOI: 10.9734/EJNFS/2020/v12i730246.

Pachkore, G. L, Dhale, D. A. (2012). Phytochemicals, vitamins and minerals content of three *Ocimum* species. *IJSID*, 2 (I): 201 - 207.

Russu, I. M., Lin, A. K., Yang, C. P. and Ho, C. (1986). Molecular Basis for the Anti-Sickling Activity of Aromatic Aminoacids and Related Compounds: A Proton Nuclear Magnetic Resonance. *Biochemistry,* 25: 808 - 815 http://dx.doi.org/ 10.1021/bi00352a012.

Shode, F. O., Koorbanally, N., Mpiana, P. T., Tshibangu, D. S. T., Oyedeji, O. O., J. D. Habila, J. D., University of Kwazulu Natal (2014). *In vitro* Anti- sickling activity of Betulinic Acid, Oleanolic acid and their derivatives, *US. Patent Apr. 1, 2014 US 8,685,469 B2,* Avalaible on http://www.google.com/patents/US8685469.

Thiraviyam, A., Sundararajan, M., Anbukkarasi, M., Thomas, P. A., Geraldine, P. (2019). A Methanolic Extract of *Ocimum basilicum* Exhibits Antioxidant Effects and Prevents Selenite-induced Cataract Formation in Cultured Lenses of Wistar Rats, *Pharmacognosy Journal*, 11(III):496 - 504.

Tshilanda, D. D., Babady, P. B., Onyamboko, V. D. N., Tshiongo, C. M., Tshibangu, T. S. D., Ngbolua, K. N., Tsalu, P. V., Mpiana, P. T. (2016). Chemo-type of essential oil of *Ocimum basilicum* L. from DR Congo and relative in vitro antioxidant potential to the polarity of crude extracts, *Asian Pacific Journal of Tropical Biomedicine,* 6(XII): 1022 - 1028. doi: 10.1016/j.apjtb.2016.08.013 b.

Tshilanda, D. D., Mutwale, K. P., Onyamboko, V. N., Babady, P. B., Tsalu, P. V., Tshibangu, D. S. T., Ngombe, N. K., Frederich, M.,

Ngbolua, K. N., Mpiana, P. T. (2016). Chemical Fingerprint and Anti-SicklingActivity of Rosmarinic Acid and Methanolic Extracts from Three Species of *Ocimum* from DR Congo, *Journal of Biosciences and Medicines*, 4: 59 - 68.

Tshilanda, D. D., Onyamboko, D. V., Tshibangu, D. S. T., Ngbolua, K. N., Tsalu, P. V., Mpiana, P. T. (2015). *In vitro* antioxidant activity of essential oil and polar and non-polar extracts of *Ocimum canum* from Mbuji-Mayi DR Congo. *J. Adv. Med. Life Sci.,* doi: 10.15297/ JALS.V3I3.04.

Tshilanda, D. D., Mpiana, P. T., Onyamboko, D. N. V., Mbala, B. M.., Ngbolua, K. N., Tshibangu, D. S. T., Bokolo, M. K., Taba, K. M., Kasonga, T. K. (2014). Antisickling activity of butyl stearate isolated from *Ocimum basilicum* (Lamiaceae). *Asian Pac. J. Trop. Biomed.,* 4(V): 393 - 398.

Vlase, L., Benedec, D., Hanganu, D., Damian, G., Csillag, I., Sevastre, B., Mot, A. C., Silaghi-Dumitrescu, R. and Tilea, I. (2014). Evaluation of Antioxidant and Antimicrobial Activities and Phenolic Profile for Hyssopus officinalis, Ocimumbasilicum and Teucrium chamaedrys. *Molecules*, 19, 5490 - 5507. http://dx.doi.org/10.3390/ molecules190 55490.

WHO (2020). *Sickle cell disease*. Available on: https://www.afro.who.int/ health-topics/sickle-cell-disease 12/10/2020.

INDEX

A

acid, x, 11, 12, 23, 25, 39, 64, 65, 76, 78, 79, 80, 81, 82, 86, 87, 88, 96, 114, 116, 117, 118, 119, 120, 122, 124, 126, 127, 128, 129, 132
aggregation, 65, 71
agricultural sector, 32
agriculture, viii, 30, 32, 43
air temperature, 59
alkaloids, vii, viii, 1, 29, 64
anthocyanin, ix, x, 52, 114, 122, 125, 126, 129
antibiotic, 14
anticancer activity, 13, 14
anticancer drug, 88
anticonvulsant, 31, 53
antioxidant, ix, x, 21, 24, 25, 27, 31, 45, 52, 53, 64, 67, 68, 69, 74, 75, 90, 91, 108, 114, 115, 118, 119, 128, 132, 133
antioxidant activity, x, 25, 67, 69, 114, 133
antisickling activity, vi, vii, x, 97, 101, 113, 114, 115, 122, 126, 129, 130, 131, 133
anti-spasmodic, 2
antitumor, 31
antiviral activity, vi, vii, x, 73, 74, 79, 85, 86, 87, 88, 89, 90, 91, 97, 98
ascorbic acid, 64, 118, 128

B

bacteria, 27, 33, 36, 37, 38, 44, 46, 48
bacterial infection, 114
beneficial effect, ix, 52, 53
biodiversity, 99
biological activities, 75
biological control, 36
biomass, 32, 35, 40, 47
blood, 54, 69, 114, 115, 116, 118, 122, 123, 126, 129
blood flow, 114
blood pressure, 54
blood transfusion, 115
blood vessels, 114
board members, 93
bone marrow, 115
bone marrow transplant, 115
brain, 15, 21, 22, 24, 66

brain damage, 15
breast cancer, 102, 104
butyl stearate, x, 114, 121, 122, 124, 126, 133

C

cancer, x, 13, 20, 24, 42, 102, 104, 114
cardiovascular disorders, 53, 64
cattle, 35, 36, 40, 43
cell lines, 13, 24
cervical cancer, 13
chemical, viii, ix, 2, 3, 13, 14, 16, 20, 24, 29, 33, 34, 36, 37, 39, 40, 43, 44, 45, 48, 52, 53, 56, 58, 75, 76, 78, 79, 81, 101, 104, 121, 122, 131
chemical properties, 16
chemical reactivity, 101, 104, 131
chlorophyll, 38, 39, 47
chromatography, 116, 117
climate, 2, 32, 54, 59
coastal region, 54, 55
composition, 3, 4, 21, 22, 23, 24, 25, 26, 27, 28, 31, 34, 35, 43, 44, 45, 58, 63, 64, 66, 68, 71, 90, 121, 122
compounds, vii, viii, ix, x, 1, 2, 13, 14, 17, 18, 19, 20, 23, 24, 25, 42, 48, 52, 53, 59, 66, 70, 75, 76, 77, 78, 79, 80, 81, 86, 90, 97, 114, 121, 122, 126, 129, 131
Congo, 73, 74, 89, 91, 95, 96, 97, 98, 99, 100, 101, 102, 103, 104, 105, 106, 107, 108, 109, 110, 111, 113, 114, 129, 130, 132, 133
constituents, 13, 14, 15, 21, 27, 31, 36, 40, 47, 70, 71, 86
consumption, 31, 129
contaminated food, 14
contamination, 14, 36, 68, 99, 102, 104
coronavirus, 74, 75, 76, 92
cosmetics, vii, viii, 1, 30, 60
COVID-19, vi, vii, ix, 50, 73, 74, 75, 76, 77, 81, 83, 84, 85, 86, 87, 88, 89, 90, 91, 95, 96, 97, 132
crop, 31, 33, 35, 37, 38, 39, 41, 44, 45, 46, 69
crop production, 45
culinary herb, 31, 42, 46, 52, 69
cultivars, ix, 3, 14, 21, 25, 49, 52, 55, 59, 61, 64, 68
cultivation, vii, viii, ix, 2, 30, 31, 39, 42, 49, 50, 52, 54, 65, 70
cytotoxicity, 24, 53, 90

D

derivatives, 31, 90, 96, 126, 132
developed countries, viii, 29
diarrhea, x, 31, 75, 114, 115
disease gene, 130
diseases, vii, ix, x, 2, 14, 19, 20, 30, 42, 52, 63, 74, 75, 88, 102, 104, 114, 129
disorder, x, 114
drugs, viii, 15, 20, 29, 75

E

economics, 41, 43, 49, 50
ecosystem, 37
electromagnetic, 16, 24
electron, 24, 121
electron microscopy, 24
energy, 33, 81, 84, 92
enterovirus, 80, 86
environment, 19, 20, 37
environmental conditions, 27
environmental factors, 92
environmental impact, 48
environmental protection, 32

extracts, x, 23, 24, 70, 71, 78, 79, 81, 86, 91, 114, 115, 116, 118, 119, 120, 122, 123, 124, 126, 127, 132, 133

F

fertilization, 34, 35, 36, 42, 45, 65, 68
fertilizers, vii, viii, 30, 31, 32, 33, 34, 35, 36, 37, 38, 41, 43, 44, 46, 49, 50
fingerprints, 111, 120
flowers, 2, 25, 31, 35, 56, 57, 86
food, viii, ix, 2, 14, 15, 20, 30, 31, 32, 52, 128
food industry, ix, 15, 52

G

gene pool, 49
gene therapy, 115
genetics, 130
genotype, 57, 58, 59, 60, 62
genus, ix, 30, 37, 52, 89, 97, 132
groundwater, 102, 104
growth, 2, 14, 16, 22, 32, 33, 34, 36, 37, 39, 40, 42, 43, 44, 45, 46, 47, 48, 49, 60, 65, 66, 69, 88

H

health, viii, ix, 14, 19, 29, 41, 47, 52, 70, 79, 95, 109, 114, 133
health care, viii, 29
heavy metals, 33, 36, 107
height, 31, 33, 34, 35, 38, 39, 40, 59, 60, 61, 62, 63
hemoglobin, x, 114, 124
hemorrhoids, x, 114, 115
human, ix, 3, 13, 14, 19, 20, 24, 30, 52, 65, 69, 70, 71, 80, 88, 103, 106
human health, ix, 14, 19, 52

human immunodeficiency virus, 80
hydrocarbons, 4, 7, 121, 122

I

immune system, 50
immunodeficiency, 80
immunomodulatory, 53
in vitro, 13, 24, 37, 64, 81, 86, 91, 92, 104, 106, 108, 132
in vivo, 27, 71, 129
industries, viii, ix, 30, 31, 52
industry, ix, 15, 52, 57
inherited disorder, x, 114
inhibition, 13, 85, 87
inhibitor, 76, 77, 90, 92
inoculation, 38, 40, 47, 48
insects, 3, 16, 17, 18, 19, 20, 53
integrated, viii, 30, 36, 49, 50, 69
integration, vii, viii, 30, 32

M

management, 31, 36, 44, 46, 49, 50, 75, 87, 95, 101, 123, 130
matter, iv, 3, 32, 33, 39, 45, 54
medical, viii, 29, 30, 94, 97, 101
medicinal, viii, ix, x, 29, 30, 31, 43, 44, 46, 48, 49, 50, 52, 53, 65, 68, 71, 74, 75, 85, 89, 90, 93, 97, 99, 100, 103, 105, 108, 110, 111, 114, 115, 130, 132
medicine, viii, ix, x, 2, 16, 29, 52, 53, 64, 74, 75, 89, 93, 95, 115, 123, 129
Mediterranean, 26, 53, 55, 58, 59, 64
Mediterranean climate, 59
metabolites, vii, 1, 2, 3, 20, 75, 92, 115
methanol, x, 114, 116, 118, 122
microorganisms, 33, 37, 38, 40
molecular docking, vi, ix, 73, 74, 75, 76, 81, 85, 88, 90, 92, 97, 131

molecules, vii, viii, ix, 29, 74, 75, 76, 81, 85, 110, 115, 122, 124, 126
Mpro SARS-CoV-2, 74
multiplication, 79, 87

N

nitrogen, 32, 34, 35, 37, 38, 39, 46, 47, 50, 60, 65, 68
nursery school, 102, 105
nutrient, 31, 36, 41, 42, 44, 49, 50, 69
nutrients, viii, 30, 32, 33, 36, 41, 45
nutritional status, 102, 105

O

Ocimum basilicum, 1, iii, v, vi, vii, viii, ix, x, 2, 3, 13, 14, 15, 17, 19, 20, 21, 22, 23, 24, 25, 26, 27, 28, 29, 30, 33, 34, 35, 36, 38, 40, 42, 43, 44, 45, 46, 47, 48, 49, 50, 64, 65, 66, 67, 68, 69, 70, 71, 73, 74, 75, 78, 85, 86, 87, 88, 89, 90, 91, 113, 114, 132, 133
oil, ix, x, 2, 3, 4, 13, 14, 15, 16, 17, 18, 19, 20, 22, 23, 24, 25, 26, 27, 28, 31, 33, 34, 35, 36, 38, 40, 41, 42, 43, 44, 45, 46, 47, 49, 50, 52, 53, 54, 56, 57, 58, 59, 60, 61, 62, 63, 64, 67, 68, 69, 71, 75, 78, 79, 81, 85, 86, 87, 88, 90, 91, 110, 114, 116, 117, 118, 121, 122, 125, 128, 129, 132, 133
oil production, 34, 42, 56
optical microscopy, 118
organic chemicals, 3
organic manures, vii, viii, 30, 32, 33, 34, 42
ornamental, ix, 2, 48, 52, 53, 55, 70
oxidative damage, 21, 22, 24
oxidative reaction, 129
oxidative stress, 15, 129

P

pathogens, 14, 19, 20, 21, 22
pharmaceutical, viii, 30, 31, 91, 97, 101
phenolic compounds, ix, 42, 52
phytomedicine, 102, 103, 105, 109
plant growth, 33, 37, 43, 47, 48, 60
plants, viii, ix, x, 3, 15, 19, 20, 26, 29, 30, 32, 33, 34, 35, 36, 37, 40, 45, 47, 49, 54, 55, 65, 68, 71, 74, 75, 85, 89, 93, 101, 109, 111, 114, 115, 123, 130, 131
protection, 3, 19, 23, 26, 32
public health, 95, 114
public interest, viii, 29

R

red blood cells, 114, 122, 126, 129
Republic of the Congo, 73, 74, 89, 99, 101, 102, 103, 104, 105, 106, 107, 108, 109, 110, 111, 113
resources, 32, 37, 89, 97, 99, 132
respiratory disorders, 75
respiratory syncytial virus, 80
root growth, 36
rosmarinic acid, x, 11, 65, 78, 80, 81, 82, 86, 91, 114, 116, 119, 120, 122, 124, 126, 127, 129, 133

S

SARS-CoV, vii, ix, 74, 75, 82, 88, 90, 97, 131
sickle cell, x, 89, 101, 114, 116, 118, 123, 126, 129, 131
sickle cell anemia, 131
species, viii, ix, 16, 17, 18, 19, 23, 29, 30, 38, 42, 52, 63, 89, 99, 100, 102, 103, 105, 119, 120, 129, 132
spectrophotometry, 118

spice, ix, x, 52, 55, 57, 60, 71, 114, 115
stress, 15, 21, 34, 36, 39, 46, 88, 129
sustainability, 33, 37, 41, 42
sustainable, v, 29, 30, 32, 37, 42, 43, 45, 46, 47, 107
sweet basil, vii, viii, ix, 1, 2, 4, 13, 14, 15, 18, 19, 21, 23, 24, 25, 30, 31, 32, 34, 41, 42, 43, 44, 45, 47, 48, 49, 50, 52, 53, 54, 55, 56, 58, 60, 63, 64, 66, 71
symptoms, 53, 75, 85
synergistic effect, 20, 21, 53
synthesis, viii, 22, 30, 43, 65

T

target, 19, 76, 77, 82
temperature, 59, 116, 117
therapeutic agents, viii, 29, 88
toxicity, 16, 75, 131

treatment, ix, x, 31, 34, 39, 42, 52, 63, 74, 75, 87, 89, 114, 115
Turkey, v, vii, ix, 1, 26, 51, 52, 53, 54, 55, 56, 58, 59, 60, 62, 64, 68, 69, 70, 71

V

variations, ix, 24, 35, 45, 52, 53
varieties, ix, 17, 27, 44, 52, 55, 58, 59, 61, 69, 86
viral infection, 84
virus replication, 79, 80
viruses, ix, 74, 81
vitamins, ix, 52, 89, 115, 132

W

worldwide, ix, x, 74, 114